Molecular Physical Chemistry
A Concise Introduction

Molecular Physical Chemistry
A Concise Introduction

K.A. McLauchlan
University of Oxford, UK

advancing the chemical sciences

ISBN 0-85404-619-4

A catalogue record for this book is available from the British Library

Published by The Royal Society of Chemistry,
Thomas Graham House, Science Park, Milton Road,
Cambridge CB4 0WF, UK

Registered Charity Number 207890

For further information see our web site at www.rsc.org

Typeset by Vision Typesetting Ltd, Manchester
Printed by TJ International Ltd, Padstow, Cornwall, UK

Preface

To write any book is an indulgence and my excuse for having done so is that I have tried to provide undergraduates with a text that differs in approach from any other I know on its subject matter. It is an attempt to provide the reader with an understanding of thermodynamics and (to a lesser extent) reactions based upon atoms and molecules and their properties rather than on one based upon the historical development of these subjects. This is possible as a result of innovative experiments performed on very small numbers of atoms and molecules that have been performed in the last decade or so.

This book makes no pretence to be a primary source of its subject matter. Rather it attempts to give molecular insight into the familiar equations of thermodynamics, for example, and should be read in conjunction with the excellent Physical Chemistry texts that already exist. It is an aid to understanding, no more and no less. But I hope that those deterred by the elegant but possibly dry approaches found in other books, which develop the subject without the properties of molecules considered, will find this more to their liking. The subjects are important to the whole understanding of Physical Chemistry and provide the underlying philosophical structure that binds its apparently separate subjects together.

I am indebted to all my students for what they have taught me and for the sheer pleasure of knowing them. But my greatest debt is to Joan, for everything important in my life.

Contents

CHAPTER 1

Some Basic Ideas and Examples

1.1 INTRODUCTION

Physical chemistry is widely perceived as a collection of largely independent topics, few of which appear straightforward. This book aims to remove this misconception by basing it securely on the atoms and molecules that constitute matter, and their properties. We shall concentrate on just two aspects and we focus mainly on thermodynamics, which although extremely powerful is one of the least popular subjects with students. A briefer account describes how reactions occur. We shall nevertheless encounter the major building blocks of physical chemistry, the foundations that, if understood, together with their inter-dependence, remove any mystique. These include statistical thermodynamics, thermodynamics and quantum theory.

The way that physical chemistry is taught today reflects the historical process by which understanding was initially obtained. One subject led to another, not necessarily with any underlying philosophical connection but largely as a result of what was possible at the time. All experiments involved very large numbers of molecules (although when thermodynamics was first formulated the existence of atoms and molecules was not generally accepted) and people attempted to decipher what happened at a molecular level from their results. This was very indirect. Nowadays the existence and properties of atoms and molecules are established and experiments can even be performed on individual atoms and molecules. This provides the opportunity for a different way of looking at the subject, building from these properties to deduce the characteristic behaviour of large collections of them, which is more in keeping with how chemistry is taught at school level. Similarly, our understanding of how reactions occur has come from observations of samples containing huge numbers of molecules and we have tried to deduce what happens at molecular level from them. Yet it is now possible to observe reactions between individual pairs of molecules, and we can reverse the procedure and start from these observations to understand

1

reactions in bulk. It is the object of this book to demonstrate the possibility of a molecular approach to thermodynamics and reaction dynamics. It is not intended as an introduction to these subjects but rather is offered as an aid to understanding them, with some prior knowledge assumed.

We start with the properties of atoms and molecules as deduced from thermodynamic measurements and from spectroscopy. This is, paradoxically, the historical approach but it establishes straightaway that the properties are directly connected to the thermodynamics and it is artificial to separate the two. But once the connection is established we show how it can be exploited to give real insight into various problems. In this chapter we introduce the fact that the energy levels of atoms and molecules are quantised and use some simple ideas to establish the effectiveness of our general approach before proceeding to their origins in the second chapter.

1.2 ENERGIES AND HEAT CAPACITIES OF ATOMS

In the gas phase, atoms move freely in space and frequently collide, at a rate that depends upon the pressure of the gas. At atmospheric pressure ($\sim 10^5$ N m^{-2}) and room temperature they move approximately 100 molecular diameters between collisions, at average velocities about equal to that of a rifle bullet (300 m s^{-1}). In elastic collisions some atoms effectively stop whilst others gain increased velocity (*cf.* collisions of billiard balls) so that instead of all the atoms having a single velocity they have a wide distribution of velocities. This is the familiar Maxwell distribution (Figure 1.1) that results from classical Newtonian mechanics. In it all velocities are possible but some are more probable than others. The most probable velocity depends upon the temperature, as does the width of the distribution.

A moving atom of mass m possesses a kinetic energy of $\frac{1}{2}mu^2$, where u is its velocity. Since in the whole collection of atoms in a gas there is no restriction to the velocity of an atom, there is no restriction to its energy either. Using the Maxwell distribution (see below), the average energy of an atom can be shown to be

$$\bar{\varepsilon} = \tfrac{1}{2}\, m\overline{u^2} = \tfrac{3}{2}kT \tag{1.1}$$

where $\overline{u^2}$ is the mean square velocity of the atoms in the sample, k is Boltzmann's constant ($k = R/N_A$ where R is the universal gas constant and N_A the Avogadro number) and T is the absolute temperature. To obtain the total energy, E, of a mole of gas we simply multiply by the total number of atoms, N_A, and obtain $(\frac{3}{2})RT$. This is the energy due to the

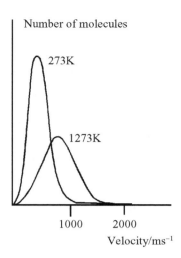

Velocity/ms^{-1}

Figure 1.1 *The Maxwell distribution of velocities of molecules in a gas at 273 and 1273 K. As the temperature is increased the most probable velocity moves to a higher value and the distribution widens, reflecting a greater range of molecular velocities.*

motion (translation) of the atoms in the gas, the 'translational energy'.

Remarkably, although the kinetic energy of an individual atom depends upon its mass, the prediction is that the total energy of the gas in the sample does not. It seemed so outrageous when first made that it had to be tested, but how? We have calculated the absolute quantity, E, but have no way of measuring it directly. But there is a closely related property that we can measure. This is the heat capacity of the system, defined as the amount of heat required to raise the temperature of a given quantity of gas (here 1 mole) by 1 K. Different values are obtained if this measurement is made keeping the volume of the gas constant (with a heat capacity defined as C_V) or keeping its pressure constant (C_P) since in the latter case energy is expended in expanding the gas against external pressure. Here we consider just what is happening to the energy of the gas itself, and must use the former. Writing the definition in mathematical form, as a partial differential (a differential with respect to just one variable, here T),

$$C_V = \left(\frac{\partial E}{\partial T}\right)_V = \left(\frac{\partial(\frac{3}{2}RT)}{\partial T}\right)_V = \frac{3}{2}R = 12.47 \text{ J K}^{-1} \text{ mol}^{-1} \tag{1.2}$$

The subscript on the bracket reminds us that we are dealing with a constant volume system.

This is again remarkable. It says that for all monatomic gases, regardless of their precise chemical nature or mass, the molar heat capacity is the same, and independent of temperature. Experiment shows this to be correct. For example, He, Ne, Ar and Kr were early shown to have precisely this value over the temperature range 173–873 K, the range then investigated.

It is now worth examining in more detail where the result that the mean energy of a monatomic gas is independent of its nature comes from. To obtain this average over the whole range of velocities we must multiply the kinetic energy of an atom at a given velocity by the probability that it has this velocity, and integrate over the whole velocity distribution normalised to the total number of atoms present. This probability function (dN/N) is the Maxwell distribution of velocities.

$$E = \tfrac{1}{2}\,\overline{mu^2} = \int\limits_{0}^{\infty} \tfrac{1}{2}\,mu^2\,\frac{dN}{N} \tag{1.3}$$

where

$$\frac{dN}{N} = 4\pi\left(\frac{m}{2\pi kT}\right)^{3/2} \exp\left(\frac{-mu^2}{2kT}\right)u^2 du \tag{1.4}$$

Thus the expression for E, clearly and understandably, contains the mass, m, and the velocity, u. Yet due to its mathematical form and since the integral is real (and is evaluated between these upper and lower limits) its value ($\tfrac{3}{2}kT$ for motion in three dimensions, Equation 1.1) does not. The expression may look formidable but the integral has a standard form (it is a *Gaussian* function) and its evaluation is straightforward; see Appendix 1.1. It follows that the same result is obtained for any form of energy (not necessarily translational) that can be expressed in the same mathematical form, $\tfrac{1}{2}ab^2$, where 'a' is a constant and 'b' is a variable that can take any value within a Maxwellian distribution. Such a term is known as a 'squared term'.

From experience, gases are homogeneous and possess the same properties in all three directions in space; for example, the pressure is the same in all directions. The motion of the gas atoms in the three perpendicular Cartesian directions is independent and we say that they have three 'translational degrees of freedom'. Resolving the velocity into these directions and using Pythagoras gives, with obvious notation,

$$u^2 = u_x^2 + u_y^2 + u_z^2 \tag{1.5}$$

with an analogous result for their means. In the gas the mean square velocities in the three directions are equal. The form of the Maxwell distribution we have used is that for motion in three dimensions and the average energy associated with each translational degree of freedom is consequently one-third of the value obtained. It then follows that for each degree of freedom *whose energy can be expressed as a squared term* we should expect an average energy of $\frac{1}{2}kT$ per atom. This is an important result of classical physics and is the quantitative statement of 'the Principle of the Equipartition of Energy'.

We stress that it has resulted from the Maxwell distribution in which there is no restriction of the translational energy that an atom (or molecule) can possess. From everyday experience this seems eminently reasonable. We can indeed make a car travel at a continuous range of velocities without restriction (and luckily personal choice of how hard we press the accelerator rather than collisions make a whole range possible if we consider a large number of cars!). But is this true of molecules that might possess other sources of energy besides translation? We should not assume so, but again put it to experimental test. We shall find later that we have to re-examine the case of translational energy too.

1.3 HEAT CAPACITIES OF DIATOMIC MOLECULES

The heat capacity, C_V, of a sample is directly related to its energy, and can be measured. We expect gaseous diatomic molecules, like atoms, to move freely in independent directions in space so that translational energy should confer upon the sample a heat capacity of $\frac{3}{2}R = 12.47$ J mol^{-1} K^{-1}. If this was the only source of energy that molecules possess then the heat capacity should have this value, and be independent of temperature. This turns out to be wrong on both counts. For example, the measured heat capacities (in J mol^{-1} K^{-1}) of dihydrogen and dichlorine at various temperatures are given in Table 1.1.

All these values are substantially greater than expected from translational motion. Through the direct relationship between C_V and E this implies that there must be additional contributions to the energy of the sample. We note also that for each gas the value increases with temperature, with a tendency for it to become constant at high temperatures for

Table 1.1 *Heat capacities (J mol^{-1} K^{-1}) of dihydrogen and dichlorines at different temperatures*

Molecule	T(K)				
	298	400	600	800	1000
H$_2$	20.52	20.87	21.01	21.30	21.89
Cl$_2$	25.53	26.49	28.29	28.89	29.10

dichlorine, and that over the whole range of temperatures in the table the heat capacity of dichlorine exceeds that of dihydrogen.

So what forms of energy can a diatomic molecule have that an atom cannot? The obvious physical difference is that in the molecule the centre of mass is no longer centred on the atoms. This implies that if there are internal motions in the molecule as it translates through space these have associated energies. The first, and most obvious, possibility is that the molecule might rotate. A sample containing rotating molecules might therefore possess both translational and rotational energy, and we need to assess the latter. The simplest, and quite good, model for molecular rotation is to treat the diatomic molecule as a rigid rotor (Figure 1.2) with the atoms as point masses (m_1 and m_2) separated from the centre of mass of the molecule by distances r_1 and r_2. Classical physics shows the rotational energy to be $\frac{1}{2}I\omega^2$, where I is its moment of inertia and ω the angular velocity (measured in rad s^{-1}). We immediately recognise this as a 'squared term'.

Rotation might occur about any of three independent axes which in general might have different moments of inertia, although for a diatomic molecule two are equal. Taking the bond as one axis (z), these are those about axes perpendicular to it through the centre of mass and their moments of inertia are defined by

$$I_x = I_y = m_1 r_1^2 + m_2 r_2^2 \tag{1.6}$$

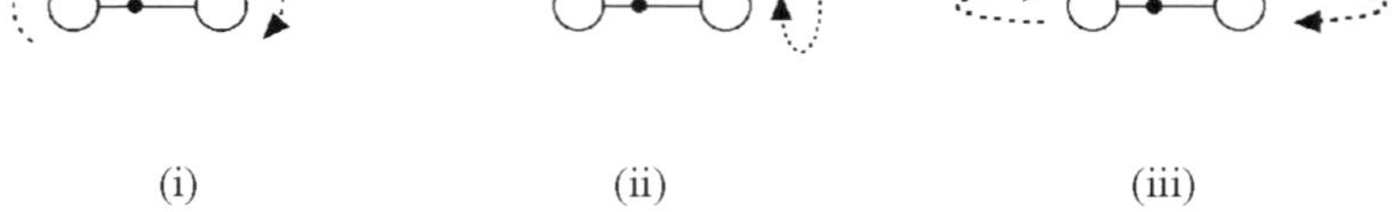

Figure 1.2 *Rotations of a diatomic molecule. The atoms are treated as point masses,* m_1 *and* m_2, *with their centres lying along the axis of the molecule with the distances* r_1 *and* r_2 *measured between these centres and the centre of mass (CM) of the molecule. The molecule can rotate about three axes, in the plane of the paper (i), about the bond axis (ii) and out of the plane (iii). The moments of inertia for rotations (i) and (iii) are non-zero and equal but the moment of inertia for rotation (ii) is zero since there is no perpendicular distance between the point masses and the CM along the bond axis. This rotation therefore does not contribute to the total rotational energy of the molecule.*

We note that the distances are measured in the direction perpendicular to the axes of rotation, here along the z-axis. This implies that the moment of inertia for rotation about the z-axis is zero because the point masses and the centre of mass all lie on a straight line, and no perpendicular distance in the x or y directions separates them. We conclude that only two of the three rotational degrees of freedom contribute to the energy of the molecule, both through squared terms in the angular velocity. Using the Equipartition Principle we predict that their contribution to the energy will be $2 \times \frac{1}{2}RT$ J mol^{-1}. This implies that, together with the translational contribution, the total energy of the molecule should be $\frac{5}{2}RT$ J mol^{-1} and C_V should be $\frac{5}{2}R$ J mol^{-1} K^{-1}. It should not vary as the temperature is changed.

This has the value 20.78 J mol^{-1} K^{-1}, which, interestingly and significantly (see later), is very close to the value observed for dihydrogen at 350 K, but Table 1.1 shows C_V to increase with temperature. However, for dichlorine it is still much too low at this temperature compared with experiment. Once again we conclude that the actual energy is greater than we thought, and that the molecule must have another form of internal motion associated with it. This is vibration.

In a vibration the atoms continuously move in and out about their average positions (Figure 1.3). As they move outwards the bond is stretched, as would be a spring, and this generates a restoring force, which

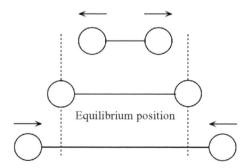

Figure 1.3 *Vibration of a diatomic molecule. The diagram shows at the top the (point mass) atoms at their distance of closest approach when they start to move apart again, in the centre at their average positions, and at the bottom when the bond is fullest stretched and the elasticity of the bond brings the atoms back towards each other once more. At the two extreme positions the atoms are momentarily stationary and the molecule possesses the potential energy obtained from stretching or compressing the bond only, but they then start to move, transforming potential energy into kinetic energy, a process complete just as the atoms pass through their equilibrium positions. If the bond stretching obeys Hooke's Law (the restoring force generated by moving away from the equilibrium position is proportional to the distance moved) then Simple Harmonic Motion results.*

if Hooke's Law is obeyed is proportional to the displacement from the equilibrium positions, and the atoms return through these positions. In this model (also quite good) the molecule behaves as a simple harmonic oscillator with continually interchanging kinetic (KE, from the motion of the atoms) and potential (PE, from stretching the bond) energies. The total energy is the sum of the two, and is conserved in an isolated gas molecule.

$$E_{\text{vib}} = (KE + PE)_{\text{vib}} \qquad (1.7)$$

At any instant the kinetic energy is given classically by $\frac{1}{2}\mu v^2$ where μ is the 'reduced mass' of the molecule (defined as $m_1 m_2/(m_1 + m_2)$) and v is the instantaneous velocity of the atoms, whilst the potential energy is $\frac{1}{2}kx^2$, where k is the bond force constant (Hooke's Law constant) and x is the instantaneous displacement from the average position of each atom. A diatomic molecule can only vibrate in one way, in the direction of the bond, but because of having to sum the contributions from both forms of energy this one degree of vibrational freedom contributes two squared terms to the total energy, through the Equipartition Principle, $2 \times \frac{1}{2}RT$ J mol^{-1}. Once again we have assumed that, in using this Principle, there are no limitations on (now) the vibrational energy that a molecule can possess.

The total energy of the molecule is, therefore, predicted to be the sum of the translational ($\frac{3}{2}RT$), rotational (RT) and vibrational (RT) contributions, giving $\frac{7}{2}RT$ J mol^{-1} and $C_V = \frac{7}{2}R = 29.1$ J mol^{-1} K^{-1}, greater than before but still independent of temperature. This is precisely the value obtained experimentally for dichlorine at 1000 K but it is much higher than that of dihydrogen at the same temperature. The heat capacities of both are still predicted, wrongly, to be independent of temperature.

It is now instructive to plot C_V against T for a diatomic molecule (shown diagrammatically in Figure 1.4). The value jumps discontinuously between the three calculated values, corresponding to translation alone, translation plus rotation and finally translation plus rotation plus vibration, over small temperature ranges (near the characteristic temperatures for rotation and vibration, θ_{rot} and θ_{vib}, Section 2.5.1). These temperatures depend on the precise gas studied, and the changes occur at higher temperatures for molecules consisting of light atoms than for those that contain heavy ones. Only at the highest temperatures are the values those predicted by Equipartition. But the contribution from translation alone is evident at temperatures close to absolute zero, but not extremely close to it when this contribution falls to zero. In this plot the translational contribution is easily recognised through its unique value but which of

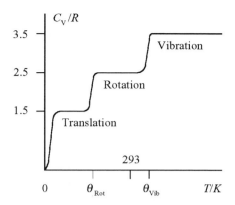

Figure 1.4 *Schematic diagram of how C_V for a diatomic molecule varies with temperature. It rises sharply from near 0 K and soon reaches 3R/2, expected for translational motion, where it remains until a higher temperature (near θ_{rot}) is reached when the rotational degrees of freedom contribute another R to the overall value. This happens well below room temperature (293 K) for all diatomic molecules. At still higher temperatures the vibrational motion eventually contributes another R, again near a rather well-defined temperature (θ_{vib}). The actual values of these temperatures vary with the precise molecule concerned, and are lower for heavy molecules (e.g. Cl_2) than light ones (e.g. H_2). The beginnings of the vibrational contribution occur below room temperature for Cl_2 but the full contribution is not apparent until well above it.*

rotation or vibration contributes at the lower temperature is obtainable only through further experiment or theory; the rotational contribution appears at the lower temperature.

We conclude that molecules exhibit very different behaviour from that we predicted using classical theory and we must examine where we might have gone wrong. All the basic equations for rotational and vibrational energy are well established in classical physics and are not assumptions. One possibility might be that the rotational and vibrational energies are correctly given classically but that they do not have Maxwellian distributions. But also we have made what in the classical world seems a wholly unexceptional assumption, *i.e.* that there are no restrictions as to the energies a molecule might possess in its different degrees of freedom. Since the predictions do not conform to the experimental observations *this might be wrong*. We speculate that rather than being able to possess *any* values of their rotational and vibrational energies, the molecule may be able to possess *only specific values* of them. This is confirmed directly by spectroscopy, see below. We describe the energies as 'quantised'. This is the basic realisation from which much of physical chemistry flows.

Translational energy seems not to be quantised but actually is; experiments have to be performed at very close to 0 K to observe this. It is why

C_V in Figure 1.4 goes to zero at this temperature. The fact that different diatomic molecules possess the two further forms of energy above characteristic temperatures that differ from one molecule to the next is an aspect of energy-quantised systems that we shall have to understand. But all systems behave classically at high enough temperatures, which may, however, be below room temperature. For example, all diatomics (save H_2) exhibit their full rotational contribution below room temperature. But only the heaviest molecules exhibit the full vibrational contribution below very high temperatures.

That the limiting classical behaviour is often observed in real systems, especially for polyatomic molecules, is ultimately why we are normally unaware of the quantised nature of the world that surrounds us. But the world is quantised in energy and we need to understand and exploit the properties of matter that this implies. Much of the new technology in everyday use depends on it.

It is fascinating and significant that this conclusion of paramount importance was indicated as a possibility through the interpretation of classical thermodynamic measurements, emphasising that a connection exists between the thermodynamics of systems and the properties of the individual atoms and molecules that comprise them.

1.4 SPECTROSCOPY AND QUANTISATION

Quantisation of energy shows itself very directly in the optical spectra of atoms and molecules and initially we consider the electronic spectra of atoms. When a sample of atoms is excited in a flame, for example, it emits radiation to yield a 'line spectrum' (Figure 1.5). This is in fact a series of images of the exit slit in a spectrometer corresponding to a series of different discrete frequencies of light emitted by the atoms. If the atom behaved classically this would not be so since the electrostatic attraction between the electron and nucleus would accelerate one towards the other, and according to classical physical laws the atom would emit light over a continuous frequency range until the electron was annihilated on encountering the nucleus. The explanation is that the energies possible for the electrons in an atom are themselves quantised, and the frequencies in the emission spectrum correspond to the electron jumping between levels of different energy, according to the Bohr condition,

$$h\nu = E_1 - E_2 \tag{1.8}$$

where h is Planck's constant, ν the frequency of the light and E_1 and E_2 are the energies of two of the levels (Figure 1.6).

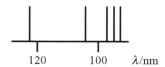

120   100   λ/nm

Figure 1.5 *Ultraviolet region of the emission spectrum of hydrogen atoms (the Lyman series). Atoms in the sample have been excited by an electric discharge to a number of higher atomic energy levels and they emit light at different discrete frequencies as the electrons return to the lowest level (principle quantum number, n = 1). This displays the quantised nature of the atomic energy levels directly. Other series of lines are observed in the visible and infrared regions of the electromagnetic spectrum, corresponding to electron jumps to different lower quantised states. The positions of the lines allow the frequencies of the emitted light to be measured.*

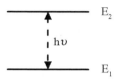

Figure 1.6 *The origin of a line in a spectrum of frequency ν. In a H atom at thermal equilibrium with its surroundings the electron is in the lower energy level and it can absorb the specific amount of energy hν to jump to a higher energy level. If, however, the atom has been excited so as to put the electron in the upper energy level then it can emit the same energy and return to the lower level. In a real atom there are many pairs of energy levels between which spectroscopic transitions can occur.*

An early triumph of the Schrödinger equation was that it rationalised why the levels are quantised and allowed the energies to be calculated, with the difference frequencies agreeing with the observed ones. Associated with each level is a mathematical function, the wave function, known as an orbital.

The C_V measurements discussed above gave no indication that monatomic gases could possess electronic energy besides translational energy. This must mean that *over the temperature range studied* the atoms do not have any. But we have seen in Figure 1.4 that different sources of energy make their contributions at different temperatures and at a high enough temperature there would indeed be an electronic contribution to the C_V of atoms. We shall see later that the crucial factor is the energy separation between the quantised energy levels compared with the 'thermal energy' (given by kT) in the system, and the lower atomic orbitals are separated from each other by large energy gaps. For some atoms, such as the halogens (see below) an electronic contribution is evident even at quite low temperatures, but this is rather unusual.

In molecules, too, discrete energy levels and molecular orbitals exist with differing electronic energies. But associated with each and every electronic level there is a set of vibrational and rotational ones (Figure 1.7). Through the Bohr condition spectroscopy allows us to measure the energy gaps directly, whilst quantum mechanics also allows us to calculate the energy levels, and the gaps between them. This confirms that the energies molecules possess are quantised. Molecular spectra are normally observed in absorption, with the molecules absorbing certain frequencies from white light flooding through a sample of them. Whereas emission spectra arise when electrons fall from higher energy levels into which the substance has been excited by heat or electricity, absorption spectra depend on species in their lower energy levels jumping to higher ones. They therefore give information on the lowest energy levels of the molecules. Experimentally a huge range of transition frequencies is involved, varying from the microwave (far infrared) to the ultraviolet regions of the electromagnetic spectrum. The former occur between energy levels that are closest in energy, those due to rotation. At higher frequencies, in the infrared, the light has sufficient energy to cause jumps between vibrational levels, but these all have more closely spaced associated rotational ones, and under certain selection rules changes occur to both the vibrational and rotational energies simultaneously. The spectra are known as 'vibration–rotation' ones (an example is given in Figure 2.6, Chapter 2). Finally, at much higher frequencies, in the ultraviolet, electrons can jump between the vibrational and rotational levels of different electronic energy states and the spectra reflect simultaneous changes in all three types of energy.

Through Equation (1.8) there is a direct relationship between frequency and energy difference and we conclude that *in terms of the gaps between the energy levels*

$$\Delta E(\text{electronic}) \gg \Delta E(\text{vibration}) > \Delta E(\text{rotation})(\text{and} \gg \Delta E(\text{translation})) \quad (1.9)$$

This confirms our conclusions from C_V measurements (Figure 1.4). At 0 K C_V is zero, but as the temperature is increased translational motion rapidly makes the contribution expected from classical physics. At a higher temperature the effects of rotation become apparent, and at a higher one still, vibration (and at very high temperatures electronic energy contributions might appear if dissociation does not take place first). *Motions corresponding to the lower energy gaps make their contributions at lower temperatures than those with higher energy gaps.*

It is important to distinguish between the absolute values of the various types of energy and the separations between the energy levels. Thus the gaps between translational levels are miniscule but even at very low temperature there is a contribution of $(\frac{3}{2})RT$ J mol^{-1} to the energy.

Energy

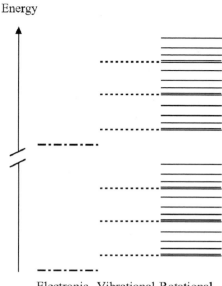

Electronic Vibrational Rotational

Figure 1.7 *Schematic diagram of the energy levels of a diatomic molecule showing only the two lowest electronic levels, which are widely separated in energy. Associated with each is a stack of more closely spaced vibrational levels, which in turn embrace a series of rotational levels. Under the Simple Harmonic oscillator approximation, the vibrational levels of each electronic level are equally spaced, but the rotational levels predicted by the rigid rotor model are not. This diagram gives an impression of the relative sizes of the electronic, vibrational and rotational energies but only a very few of the lower energy levels can be shown here without loss of clarity. Real molecules have far more levels. Spectroscopic transitions can be excited between the rotational levels alone, between the rotational levels in different vibrational states and between the sub-levels of the electronic energy states. The three types occur with very different energies and are observed in quite different regions of the electromagnetic spectrum.*

The gaps between electronic energy levels are enormous in comparison, yet electronic energy makes no contribution to the total energy of most systems at room temperature. A common error is to believe that since the energies of molecular electronic levels may be high then the electronic energy of the system must be high. If the molecule existed in one of the higher levels the energy would indeed be high. But it does not at low temperatures, where the molecule is in its lowest electronic level. For a gas at room temperature the translational energy is greatest in magnitude followed by the rotational energy and then the vibrational one. The order is exactly reversed from that of the energy gaps:

$$E(\text{translational}) > E(\text{rotational}) > E(\text{vibrational}) \, (\gg E(\text{electronic})) \quad (1.10)$$

1.5 SUMMARY

Atoms and molecules may possess several different contributions to their total energy but each one can have only certain discrete quantised values. The energies associated with different modes of motion differ in magnitude, with the gaps between the energy levels for translation being smaller than those for rotation that, in turn, are smaller than those for vibration. The gaps between the electronic energy levels of molecules normally vastly exceed all of these. This causes translational motion to occur at lower temperatures than rotational motion, which occurs at lower temperatures than vibration (and much lower than electronic excitation). Yet we remain unaware of quantisation in every day life, and we have some indication already that this is inherently because systems behave classically at high enough temperatures. We need to sharpen this concept and decide what a 'high enough' temperature is, and then we should be able to predict the behaviour of systems from a knowledge of the quantised energy levels of the atoms and molecules from which they are made.

1.6 FURTHER IMPLICATIONS FROM SPECTROSCOPY

We are so familiar today with spectra that we tend to miss a very remarkable fact about them. An atomic spectroscopic transition in absorption, for example, occurs when an atom in a specific energy level accepts energy from the radiation and jumps to another specific level, in accordance with the Bohr condition and under various selection rules that limit the transitions that are possible. These have been discovered by experiment, and can be rationalised using quantum mechanics. So the spectrum of a single atom undergoing a single transition is a single line at one specific frequency. But when we talk about the spectrum of an atom (a loose term) we immediately think of a whole family of transitions at different frequencies (Figure 1.5). Since the electron in one atom can only make one jump between energy levels at a time it follows that what we see is the result of a *large number* of individual atoms simultaneously absorbing energy and jumping to a whole range of possible quantum states. That is, instead of seeing the spectrum of a single atom, we are observing the spectra of a very large number of individual atoms simultaneously. We say that spectroscopy is an *ensemble phenomenon*, meaning precisely that what we observe is the result of what is happening in the large collection of atoms.

But what would happen if we were clever enough to observe a single atom over a long period of time, rather than instantaneously? Following the initial absorption of energy from the light beam, the atom enters a higher energy level. Let us now postulate that an efficient mechanism exists (it does!) for returning it to the lowest level, where it might absorb a different frequency from the incident radiation and attain a second, different, higher level. It would then return and the process be repeated over a whole cycle of possible transitions covering the total range of frequencies. We conclude that over infinite time the atom would perform all the possible transitions allowed to it and the spectrum of the single atom would be identical in appearance to the ensemble one. This has been put to direct experimental test in recent years, although in molecular rather than atomic spectroscopy, and found to be correct. It opens the possibility of calculating ensemble behaviour of a collection of molecules from the behaviour over time of a single one, a concept close to the basis of using ensembles in statistical thermodynamics (see later).

Another unexpected aspect of spectroscopy lies in the Bohr condition. We tend to think of atoms or molecules absorbing energy from an incident light beam and jumping to higher energy states, but Equation (1.8) does not dictate a direction for the energy change to occur. That is, whilst an atom in a lower energy state might jump to a higher energy state, one in that state might emit energy under the influence of the light beam, and fall back to the lower state. In this case the beam would exit more intense than it arrived. These processes are known as *stimulated absorption* and *stimulated emission* respectively. Einstein was the first to consider this and showed by a simple kinetic argument (it is now more satisfyingly done using quantum mechanics) that the absolute probability of an upward or a downward transition caused by light of the correct frequency is exactly the same. It follows that if there are molecules in both energy levels then the intensity of a spectroscopic line depends on the difference between the number of atoms that absorb energy from the light beam and those that emit energy to it, and therefore on the *difference* in the populations of the two energy levels. This is further considered in Section 4.4. So not only can spectroscopy measure the energy gaps in atoms and molecules, but it indicates this difference in populations, too.

Since all atoms and molecules at thermal equilibrium with their surroundings are found experimentally to exhibit absorption spectra, we conclude that *at thermal equilibrium* the lower energy states are the more highly populated. Samples in which the atoms are deliberately excited to higher levels, for example, by an electric discharge through them, have their upper levels overpopulated and exhibit emission spectra. This is how streets are lighted, using the emission spectrum of sodium. We should always remember that systems are not necessarily at thermal equilibrium; indeed, equilibrium can be disturbed or avoided in many ways of increasing technical importance. For example, lasers depend on deliberately producing overpopulations of higher energy states.

1.7   NATURE OF QUANTISED SYSTEMS

The quantised world is a strange and un-instinctive one, complicated by the fact that polyatomic molecules possess millions of discrete energy levels. But we can understand what quantisation implies by studying it at its simplest and we initially consider a system that has just two energy levels available to it. That is, one in which the energy cannot vary without restriction, as in the classical physical world, but in which the atoms, molecules, nuclei or whatever that comprise the system can individually adopt one of just two possible energy states. Such systems actually exist. For example, the nucleus of the hydrogen atom, the proton, is magnetic and can interact with an applied magnetic field to affect its energy. Experiment shows that it is a quantum species whose magnetic moment can adopt either of two orientations with respect to the field direction, rather than the one that a classical compass needle would. In one orientation the magnetic moment lies along the direction of the applied field, and its energy is lowered, whilst in the other it opposes it, and its energy is increased. These are simple experimental facts and they imply that application of the field creates a two-level system [Figure 1.8(i)]. This is exploited in Nuclear Magnetic Resonance (NMR) spectroscopy, in which transitions are excited between the two levels. A different example is found in the halogen atoms that behave as though they were two-level systems at low temperature.

Let us be clear what is implied by the existence of the two levels. We define the energy of the lower to be 0, and that of the upper one ε. If we have one particle (an atom, a nucleus or whatever) in its lowest level its energy is 0, whereas if, somehow, we put it into the upper state its energy is ε. It is crucial to our understanding of quantised systems that *there is no other possibility*. The particle cannot, for example, have an energy of $\varepsilon/2$ or 1.2ε. This seems at odds with everyday life in which, up to some limit, systems seem to be able to possess any energy. For example, we can heat a kettle to any temperature below the boiling point of water. We need somehow to reconcile this difference with the classical world since we know that at atomic or molecular level *all* energies are quantised.

The secret once more lies in the fact that in the experiments we usually perform we do not study individual particles but rather collections of them in which they may be distributed between the two energy states. We start by simply adding a second [Figure 1.8(ii)]. Now both may be in the same energy level, to give a total energy of 0 or 2ε or one may be in one

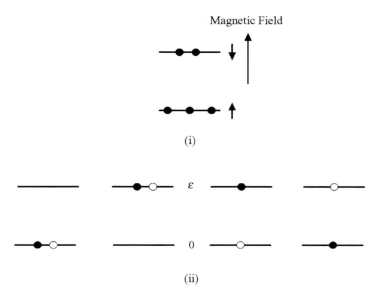

(i)

(ii)

Figure 1.8 (i) *A two-level system results when hydrogen nuclei are placed inside a magnetic field. Some protons align with their magnetic moments along the applied field, and some against it, resulting in two different energy states. At thermal equilibrium there are more nuclei in the lower energy state than in the upper one. Nuclear Magnetic Resonance Spectroscopy consists in causing a transition between the two. (ii) If just two particles enter a two-level system their total energies can be 0, 2ε or ε, but if they enter the levels with equal probability then the latter can be obtained in two ways, unlike 0 and 2ε which can be obtained in just one.*

level and the other in the other, giving a total of ε. Now consider the *average* energy of the two. For this latter case this is ε/2, so that we now have an energy that is not one of the quantised values. The *total* energy is simply the number of particles times the average value (2 × ε/2), and in this case has a value equal to that of one of the quantised levels. But this is rarely true as we increase the number of particles in the system. Consider one in which there are $(m + n)$ particles divided between the two energy states, with m in the lower energy one. Now the total energy is $n\varepsilon$, which can take a range of values determined by n, and the average energy

$$\bar{E} = \frac{n\varepsilon}{(m + n)} \tag{1.11}$$

which is determined by the values of m and n. Whereas the energies of the individual particles have discrete, quantised, magnitudes, the average energy, and the total energy, have no such restrictions and can take a wide range of values. This lies at the heart of why systems containing species with quantised energy levels nevertheless display classical behaviour under most conditions. These energies clearly depend on how the particles are distributed over the two energy states, that is on the numbers m and n.

This simple example leads to a further important insight. If we can distinguish between the particles, *i.e.* know which is which, then the situations with 0 or 2ε in total energy can each be obtained in just one way. But an energy of ε is obtained if either is in the lower level and the other in the upper one, in two ways. If both levels are equally likely to be populated this energy is twice as likely to occur as either of the others. Generalising, this implies that some population distributions and some energy values are more likely to occur than others. This is a conclusion of momentous importance. However, we stress that it depends on the likelihood of populating each level being inherently equal.

We shall now consider systems containing a large number of, *e.g.* molecules, which may exist in any of a large number of energy levels, before returning to the two-level system.

1.7.1 Boltzmann Distribution

Within chemistry we habitually deal with systems that contain a very large number (N) of molecules and we consider how these are distributed between their numerous energy levels when the systems are in thermal equilibrium with their surroundings at temperature T. For example, 1 mole of gas at 1 bar pressure contains N_A (6.022×10^{23}) molecules. We need the quantum analogue of the Maxwell distribution of energies. It is given by the Boltzmann distribution, which we shall state and use here before deriving it in Chapter 2. It is a statistical law that applies to a constant number of independent non-interacting molecules in a fixed volume, and is subject to the total energy of the system being constant (the system is isolated), and the several ways of obtaining this energy by distributing the molecules between the quantised levels being equally likely. That is, it does not matter which particular molecules are in specific energy states provided that the total energy is constant. The distribution is

$$\frac{n_i}{N} = \frac{g_i e^{-\varepsilon_i/kT}}{\sum_i g_i e^{-\varepsilon_i/kT}} \tag{1.12}$$

where n_i is the number of molecules in a level of energy ε_i and g_i is the degeneracy of that level (the number of states of equal energy, for example, the three p orbitals of a H atom have the same energy and so for this $g_i = 3$). The denominator includes summation over all the energy levels of the molecules. This equation results from statistical theory applied to a very large number of molecules, under which conditions one particular distribution of the molecules between the quantised levels becomes so much more likely than the rest that it alone need be considered. This is the ultimate extension of our conclusion concerning just two levels.

On first encounter, the Boltzmann distribution looks formidable, especially because it apparently involves summation over the millions of energy levels present in polyatomic molecules. But we now return to the two-level system to discover the circumstances where this is only an apparent difficulty.

1.7.2 Two-level Systems

Consider a two-level system in which there are n_0 particles (atoms, molecules or whatever) in the lower level and n_1 in the upper one so that the total number $N = (n_0 + n_1)$. At thermal equilibrium the Boltzmann distribution tells us that these are related through

$$\frac{n_1}{(n_0 + n_1)} = \frac{g_1 e^{-\varepsilon/kT}}{g_0 e^{-0/kT} + g_1 e^{-\varepsilon/kT}} = \frac{g_1 e^{-\varepsilon/kT}}{g_0 + g_1 e^{-\varepsilon/kT}} \qquad (1.13)$$

Rearrangement yields

$$\frac{n_1}{n_0} = \frac{g_1}{g_0} e^{-\varepsilon/kT} \qquad (1.14)$$

In the simplest case the degeneracy of each state is 1, *e.g.* for protons inside a magnetic field. Here the ratio of the populations depends directly and solely on the value of the dimensionless exponent (ε/kT) that varies as the temperature is changed, ε being a fixed characteristic of the system. The denominator, kT, is known as the 'thermal energy' of the system. This energy is always freely available to us in systems at thermal equilibrium with their surroundings and, indeed, we cannot avoid it without decreasing the temperature to 0 K. This gives us a simple physical picture. The thermal energy is what a system possesses by virtue of the motion of the particles that comprise it, and we see that it is closely related, for example, to the mean thermal energy due to translation $\frac{3}{2}kT$, above. But the distribution tells us that in quantised systems we must compare kT to ε rather than $\frac{3}{2}$ times it. If kT is much less than ε we do not have the energy to raise the particle from its lower energy level to the higher one, but as the temperature is increased it becomes possible to do so.

So much for the basic picture; now let us investigate the distribution semi-quantitatively. At low temperatures $kT \ll \varepsilon$ and for the sake of argument we let $kT = 10^{-4} \varepsilon$ so that $(\varepsilon/kT) = 10^4$. Now,

$$\frac{n_1}{n_0} = e^{-10000} \approx 0 \tag{1.15}$$

As expected there are essentially no particles in the upper level. Now we increase the temperature to make $kT = \varepsilon$ so that $(\varepsilon/kT) = 1$, giving

$$\frac{n_1}{n_0} = e^{-1} \approx 0.37 \tag{1.16}$$

and now many of the particles are in the upper state, although interestingly not all of them are despite us seemingly having enough energy to put them there. This reflects the statistical nature of the distribution and is most easily thought of in terms of the classical translational energy described by the Maxwell distribution: the *average* energy may equal the energy gap but the atoms possess a range of energies, not all of which are suitable for providing the precise value for the quantum jump.

However, we have started to populate the upper level significantly and our classical instinct would be that as the temperature is raised further we would eventually give all particles this energy, and all would be in the upper state. But the Boltzmann distribution tells us that this is completely wrong, as can be seen by increasing T until $(\varepsilon/kT) = 10^{-4}$ say, when

$$\frac{n_1}{n_0} = e^{-0.0001} \approx e^0 = 1 \tag{1.17}$$

That is, no matter how high we raise the temperature the most we can accomplish in this two-level singly-degenerate system is to equalise the populations of the two states, with $n_1 = n_0$. To an extent this reflects the fact that once kT exceeds ε an individual atom, for example, cannot accept energy since, as we have stressed above, it can only possess 0 or ε. The full variation of the number of particles in the upper state as the temperature is changed is shown in Figure 1.9(i). The population of the upper level increases exponentially from zero at 0 K but tends to the asymptotic value of $N/2$ at high temperature.

The energy of this system can now be calculated very simply since at any temperature n_0 particles have zero energy and n_1, ε:

$$E = n_0 \times 0 + n_1 \times \varepsilon = n_1\varepsilon \tag{1.18}$$

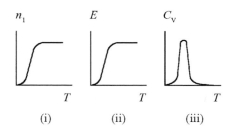

Figure 1.9 (i) *In a two-level system all the population is in the lower level at absolute zero but the number in the upper level (n_1) grows as the temperature is increased. It does not grow indefinitely, however, but reaches an asymptotic value at high temperature. If the levels are singly-degenerate, half of the total number of molecules is then in each level. (ii) The total energy of the system is the product of the population of the upper level times its energy (see text) and so it varies with temperature exactly as does n_1. The energy also reaches an asymptotic value at high temperature. (iii) Variation of C_V with temperature is given by the differential with respect to temperature of the energy curve, (ii) It starts from zero, goes through a maximum at the point of inflexion of the energy curve, and returns to zero at high temperature.*

This is a simple multiple of n_1 so that the variation of energy with temperature has exactly the same form as the variation of n_1 [Figure 1.9(ii)]. This is astonishing. It shows that as the temperature is increased E does not increase continually but again tends to an asymptote. To check this we must again turn to a calculation and measurement of C_V, which is obtained over the temperature range simply by differentiating Figure 1.9(ii). $[C_V = (\partial E/\partial T)_V]$. This is shown in Figure 1.9(iii). It predicts, what would be very strange in classical physics, that the heat capacity increases through a maximum and then falls to zero as the temperature is increased, and is what is observed experimentally.

This precise behaviour is unique to the two-level system. But the arguments we have used are not. Thus the Boltzmann distribution in any system, in which any particle may exist in any one of a number of discrete energy levels, always contains exponential terms in which quantised energies are compared with kT, and it is the values of these exponentials that largely determine level populations and all the physical properties of the sample. That is, they all depend upon the ratio (ε/kT). This simple realisation gives probably the most important insight into physical chemistry.

An example is seen if we extend the argument used above to calculate the energy of the two-level system to one with many levels. We realise that the particles are distributed amongst the levels each of which has its own characteristic quantised energy, ε_i, and we must sum the energies of them all. It follows that

$$E = \sum_i n_i \varepsilon_i \tag{1.19}$$

This seemingly obvious statement has astounding implications when we realise what we have done. It says that if we know the values of the quantised energies, which we can determine from spectroscopy or calculate using quantum mechanics, then we can calculate a thermodynamic property. It establishes a direct relationship between the properties of individual atoms and molecules and the thermodynamic properties of samples made up of large numbers of them, and it is one of the fundamental equations of statistical thermodynamics. We shall return to it later.

1.7.3 Two-level Systems with Degeneracies Greater than Unity; Halogen Atoms

At room temperature the electrons in atoms are found exclusively in their lowest orbitals. This is because the higher orbitals are greatly separated in energy from them so that $(\varepsilon/kT) \gg 1$. This is true of the halogens, but with these the lowest level is split into two by spin–orbit coupling, which is smallest in fluorine and largest in iodine (Figure 1.10). Such coupling results because motion of the electrically charged electron around the nucleus in a p-orbital causes a magnetic field there that is experienced by the electron itself. However, the electron possesses spin angular momentum that causes it to have a quite separate magnetic moment. As with the proton in an external magnetic field the quantised electron magnetic moment can adopt just two orientations inside the field due to orbital motion, and two energy levels result. Experiment shows that the energy separation between them is very low compared with the energies separating the orbitals so that the halogens behave as if they were two-level systems at normal temperatures.

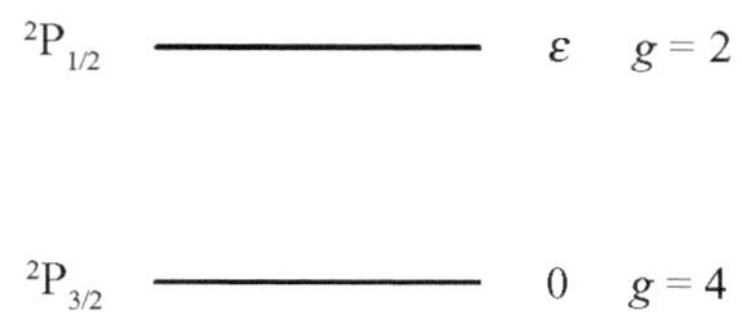

$$^2P_{1/2} \quad \underline{\hspace{3cm}} \quad \varepsilon \quad g = 2$$

$$^2P_{3/2} \quad \underline{\hspace{3cm}} \quad 0 \quad g = 4$$

Figure 1.10 *The lowest electronic energy level of the halogen atoms is split into two by spin–orbit coupling, the interaction between the magnetic moments due firstly to the orbital motion of the electron in its p-orbital and secondly due to its intrinsic spin. In fluorine the splitting in energy is of the order of kT at room temperature and both levels are populated. The next lowest electronic level is comparatively very high in energy (energy $\gg kT$) and is completely unpopulated at room temperature. At this temperature the atoms behave as though they are two-level systems, but the two levels have different degeneracies.*

We take F as an example. Its electron configuration is $1s^2 2s^2 2p^5$ so that it has a single unpaired electron in a p-orbital. The states that result from spin–orbit coupling can be calculated simply using Term Symbols (Appendix 1.2) that show the ground state to split into $^2P_{1/2}$ and $^2P_{3/2}$ components. The subscript indicates the total angular momentum quantum number of the state, J, and according to Hund's rule the state with the highest J, $\frac{3}{2}$, lies lowest in energy when an electron shell is over half full. However, the Term Symbol has another crucial piece of information encoded in it since the degeneracy of a state is $(2J + 1)$, and we have seen that the degeneracy enters the Boltzmann distribution. For the upper state $g_1 = 2$ whilst for the lower one $g_0 = 4$ (Figure 1.10) so that

$$\frac{n_1}{n_0} = \frac{2}{4}e^{-\varepsilon/kT} \tag{1.20}$$

The exponential term changes with temperature exactly as before, tending to unity as $T \to \infty$. In consequence the asymptotic value of the ratio is no longer 1 but 0.5. That is, at high temperatures one-third of the atoms are in the upper state compared with the half obtained when the levels have equal degeneracies. Had the state with $J = \frac{3}{2}$ been the upper one then two-thirds of the atoms would have been in the upper state at the higher temperatures. Simply put, at high temperatures, a state of degeneracy g can hold g times more atoms than one of degeneracy one – the states behave as though they were buckets. Degeneracy has a significant effect on level populations.

Through the direct relationship between energy and heat capacity it is clear that the halogen atoms have C_V values that reflect their ability to accept electronic energy within these split ground state levels besides possessing translational energy.

It remains to put in some values to see how significant this is. Notably, the exponent always appears as a ratio so, provided that we express numerator and denominator in the same units, it does not matter what the units are. Spectroscopists measure ε using experimentally convenient reciprocal wavelength units, denoted $\bar{v}$ and usually quoted in cm^{-1} (1 cm $= 10^{-2}$ m). These are directly related to energy through the relations $E = hv$ and $c = v/\lambda$ or $c = v\bar{v}$, where v, λ and c are respectively the frequency, wavelength and velocity of the light. It follows that $E = hc/\bar{v}$. But rather than calculating this each time it is convenient to calculate kT/hc in cm^{-1} and to use the measurement units. (A discussion on units is provided in Appendix 2.1, Chapter 2). For F, $\varepsilon = 401$ cm^{-1}, whilst kT/hc at 298 K (room temperature) $= 207.2$ cm^{-1}, so that $\varepsilon/kT = 1.935$, and $e^{-\varepsilon/kT} = 0.144$. To work out the electronic contribution to the energy of 1 mole of F atoms at this temperature we first return to the Boltzmann distribution to obtain the number of atoms in the upper state:

$$\frac{n_1}{N_A} = \frac{g_1 e^{-\varepsilon/kT}}{g_0 e^{-0/kT} + g_1 e^{-\varepsilon/kT}} = \frac{2 \times 0.144}{4 + 2 \times 0.144} = 0.067 \qquad (1.21)$$

where, as usual, we have defined the lower state to have zero energy. The total electronic energy per mole at this temperature is $E = n_1 \varepsilon N_A = 0.067 \times 401 N_A$ cm^{-1} mol^{-1}, which is 328 J mol^{-1}. Thus the presence of the low-lying $^2P_{1/2}$ state increases the total energy of the system from the pure translation value of $12.47 \times 298 = 3716$ J mol^{-1} (recall $C_V = 12.47$ J mol^{-1} K^{-1} for translation) by roughly 9%. This increases rapidly with temperature.

The electronic heat capacity is given by $C_V = (dE/dT) = \varepsilon(dn_1/dT)$, where the latter is obtained by differentiating the previous equation.

1.7.4   A Molecular Example: NO Gas

NO is the only simple diatomic molecule that contains a single unpaired electron and it is in a π^* orbital, implying that it possesses one unit of orbital angular momentum about the bond axis, yielding a magnetic moment. Magnetic interaction with the electron spin magnetic moment once more results in spin–orbit coupling (Figure 1.11) and the ground state is split into two, giving $^2\Pi_{1/2}$ and $^2\Pi_{3/2}$ states, with now the former the lower in energy. In diatomic molecules, as opposed to atoms, the degeneracies cannot be assessed from thesesymbols, but each is doubly degenerate ($g_0 = g_1 = 2$). The higher state lies 121 cm^{-1} above the lower so that at room temperature $\varepsilon/kT \approx 0.58$ and

$$\frac{n_1}{n_0} = \frac{2}{2} e^{-\varepsilon/kT} \approx 0.56 \qquad (1.22)$$

showing that over one-third of the molecules are in it at room temperature. As before, we could work out what this implies as a molar contribution to the total energy of the system but it is obviously appreciable, and so is the effect on the heat capacity. But, as with all light diatomic molecules, there is no contribution to the heat capacity at this temperature from vibrational motion, since the first excited vibrational level is too far removed in energy from the ground state to be occupied.

Clearly, it is simple to calculate the electronic energy and heat capacity of F and NO. But our calculations need not be restricted to these properties. A sample of NO at room temperature is found to be magnetic

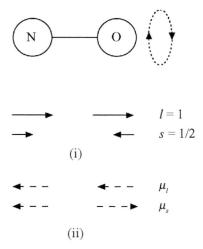

Figure 1.11 *In NO there is an unpaired electron in a π-orbital with orbital angular momentum (with quantum number l = 1) about the bond axis; the vector representing this is therefore drawn along this axis. (i) Relative to this the spin angular momentum vector of the electron (s = ½) can lie either parallel to it, to give an overall momentum of ³⁄₂, or antiparallel to it ½, leading to $^2\Pi_{3/2}$ and $^2\Pi_{1/2}$ states. (ii) These motions of the negatively charged electron produce magnetic moments, also along the axis but in the opposite direction to the angular momentum vectors, and experiment shows that the magnetic moment due to spin motion is (almost) equal to that due to orbital motion. In the former state the moments add to make the state magnetic but in the latter the moments are opposed and the state is not magnetic. In the molecule, as opposed to the atom, each state is doubly degenerate (g = 2) and, with the next lowest molecular orbital well removed in energy from the lowest, NO behaves as a two-level system exactly as shown in Figure 1.9.*

(actually paramagnetic). At first sight this seems unsurprising in a molecule that contains an unpaired electron, since electrons are magnetic, but we must remember that in spin–orbit coupling we have discovered the influence of a second magnetic field within the molecule, due to orbital motion. We have therefore to consider the resultant magnetic field of the two rather than just that of electron spin. By quantum laws these can lie only parallel (the $^2\Pi_{3/2}$ state) or antiparallel (the $^2\Pi_{1/2}$ state) to each other along the molecular axis. In the former case the magnetic moments re-enforce each other, and in the latter they are opposed. It was discovered experimentally that the magnetic moment due to the spin motion is almost exactly equal (to 0.11%) to that due to orbital motion so that the two cancel in the $^2\Pi_{1/2}$ state, making it essentially non-magnetic. Since this is the lower energy state all the molecules would be in it at sufficiently low temperature and the sample would not be appreciably magnetic. That a room temperature sample is magnetic results from thermal population of the upper state.

A measure of the magnetism of a bulk sample is its susceptibility (χ, Greek chi) which is again straightforward to calculate. Calling the magnetic moment of a molecule in the $^2\Pi_{3/2}$ state μ, it is given by

$$\chi = n_{1/2} \times 0 + n_{3/2} \times \mu = n_{3/2}\mu \tag{1.23}$$

and $n_{3/2}$ per mole is obtained from the Boltzmann distribution as before (but remember $g_0 = g_1$ here). The temperature dependence of the susceptibility has exactly the same mathematical form as does the population of the upper state, and the energy of the system. It therefore increases from (near) zero at 0 K and rises to an asymptotic value. We see how powerful some rather straightforward ideas are in calculating the physical properties of collections of atoms and molecules.

APPENDIX 1.1   THE EQUIPARTITION INTEGRAL

From Equations (1.3) and (1.4),

$$\tfrac{1}{2} m\overline{u^2} = 2\pi m \left(\frac{m}{2\pi kT}\right)^{3/2} \int_0^\infty e^{-mu^2/2kT} u^4 du \tag{1.24}$$

Since the standard integral

$$\int_0^\infty e^{-ax^2} x^4 dx = \frac{3\pi^{1/2}}{8a^{5/2}} \tag{1.25}$$

we find that the terms in m cancel and

$$\tfrac{1}{2}m\overline{u^2} = \tfrac{3}{2}kT \tag{1.26}$$

This result obviously holds for any 'squared term' whose energy distribution is given by the Maxwell equation.

APPENDIX 1.2   TERM SYMBOLS

A Term Symbol provides a straightforward means for assessing what states of an atom exist as a result of spin–orbit coupling from knowledge of the electronic structure of the atom. In general an atom possesses many electrons and coupling occurs between their spin and orbital motions according to definite rules. For light atoms (those with low atomic numbers) *Russell–Saunders* coupling decrees that the spin and orbital angular momenta sum according to the rules:

$$S = \sum_i s_i \tag{1.27}$$

$$L = \sum_i l_i \tag{1.28}$$

where S is the vector sum of all the individual spin vectors of the electrons, s_i, and likewise for L; the quantum unit of angular momentum $(h/2\pi)$ is omitted by convention. The total angular momentum resulting from coupling between the magnetic moments due to these resultant momenta is then given by

$$J = L + S \tag{1.29}$$

A further convention is to write all these quantities in terms of the quantum numbers (scalars) rather than actual values of the angular momenta, despite having to remember that *vector* addition is involved. The result is expressed in the Term Symbol

$$^{2S+1}L_J \tag{1.30}$$

Orbital angular momentum quantum numbers of individual electrons are given letter symbols. Thus s, p, d and f refer to l values of 0, 1, 2 and 3 respectively. When the vector addition has taken place capital S, P, D and F symbols are used for the corresponding total L values.

In filled electron shells individual s_i values cancel, as do the corresponding l_i ones, and we need consider only the unpaired electrons. In the F atom, with one unpaired electron in a 2p-orbital, $S = s_i = \frac{1}{2}$ and $L = l_i = 1$, which is a P state. By the general laws of quantum mechanics the two vectors can lie only parallel or antiparallel to each other, yielding just two possible values of J, $\frac{3}{2}$ and $\frac{1}{2}$. Their Term Symbols are $^2P_{3/2}$ and $^2P_{1/2}$ respectively, since $(2S + 1) = 2$. They are described as 'doublet P three halves' and 'doublet P half' states. Their energy separation depends on the spin–orbit coupling constant, which differs between different halogen atoms. From general quantum mechanical principles, the degeneracy of each state is given by $(2J + 1)$.

With diatomic molecules a similar convention is used, except that the total angular momenta are summarised in Greek alphabet symbols, Σ (sigma), Π (pi), and Δ (delta) corresponding to S, P and D in atoms. The basic symbol becomes $^{2\Sigma+1}\Lambda$ where Σ is analogous to S and Λ (lambda) to L, and spin–orbit coupling may again occur between the two momenta. For NO with its single unpaired electron in a π^* orbital ($\Lambda = 1$, a Π state) the resultant states are, consequently, $^2\Pi_{1/2}$ and $^2\Pi_{3/2}$. But these both have degeneracies of 2, not what we would expect for atoms. They are referred to as 'doublet Pi half' and 'doublet Pi three halves' states.

PROBLEMS

1.1   A linear molecule containing n atoms has 3 degrees of translational freedom, 2 degrees of rotational freedom and $(3n-5)$ degrees of vibrational freedom that contribute to its heat capacity. A non-linear one has 3, 3 and $(3n-6)$ degrees respectively.

Use the Equipartition Theorem to predict the molar internal energies and heat capacities of (i) O=C=O (linear) and (ii) H_2O (bent).

What would determine whether these values would be observed at 298 K?

1.2   In the nuclear magnetic resonance (NMR) experiment the degeneracy of the spin states of a spin-$\frac{1}{2}$ nucleus is removed by the application of an external magnetic field, B, measured in Tesla, T [see Figure 1.8(i)]. The energies of the resultant states are given by $m_I\mu B$ where $m_I = \pm\frac{1}{2}$ and μ is the magnetic moment; for the ^1H nucleus its value is 2.44×10^{-26} J T^{-1}. For 1 mole of H atoms in an external field of 14 T calculate

(i) the populations of the two spin states at 298 K. Why is it advantageous to perform the NMR experiment in the highest external field possible?
(ii) The contribution to the energy.
(iii) The magnetic susceptibility [$= (n_{\text{lower state}} - n_{\text{upper state}})\mu$].
(iv) The transition frequency between the two states in the NMR experiment.

1.3   The electron is also a spin-$\frac{1}{2}$ particle, but its magnetic moment is -1.61×10^{-23} J T^{-1}. Electron spin resonance (ESR) experiments consist of exciting transitions between energy levels made non-degenerate by applying an external magnetic field. The negative sign makes the $m_S = +\frac{1}{2}$ state the lower in energy. If this experiment could also be performed in a 14 T field (it cannot at present for technical reasons) calculate the relative populations of the two energy levels. Assuming that the probability of causing a transition is the same in each case, what are the relative sensitivities of the ESR and NMR experiments?

1.4   The electronic ground state of F atoms is split by spin–orbit coupling as described in the text. The upper level is 401 cm^{-1} above the lower level. Calculate the population of the upper level of 1 mole of F atoms at (i) 298, (ii) 798 and (iii) 1298 K and the internal energy at these temperatures.

1.5   In Cl the spin–orbit splitting is 882.4 cm^{-1} and in Br it is 3685.0 cm^{-1}. The next lowest level in each case is at 71 958.4 and 63 436.5 cm^{-1}

respectively. Calculate the molar populations at 298 K of the upper of the two lowest states for both atoms in terms of N_A, and the electronic contribution to the energy per mole.

1.6   The lowest electronic energy level of O atoms is split into three by spin–orbit coupling. The Term Symbols of the states produced are $^3P_2$, $^3P_1$ and $^3P_0$, with energies of 0.00, 158.3 and 227.0 cm^{-1} respectively. What are the degeneracies of the three states? Use Equation (1.11) to calculate their relative populations at 298 K and then Equations (1.11) and (1.18) to calculate the electronic energy of 1 mole of O atoms. The next lowest level is at 15 867.9 cm^{-1}.

1.7   Alkali metals after Li have (core)np^1 electron configurations and therefore have ^2P ground states that are split by spin–orbit coupling. For Na, K, Rb and Cs the splittings are, respectively, 17.2, 57.7, 238 and 554 cm^{-1}. Assuming that in each case the next lowest level is so high in energy that all these metals behave as two-level systems, calculate the molar populations at 298 K of the levels in each in terms of N_A.

   For Cs, how close in energy would the next electronic level have to be for it to have a population approximately equal to 1% of that of the higher state? (Assume $g_{upper} = 1$ and that any change in the population of the $^2P_{3/2}$ is negligible; note that in these atoms the outermost shell is less than half-filled and so the $^2P_{1/2}$ state is the lower in energy.)

CHAPTER 2

Partition Functions

Chapter 1 established that atoms and molecules possess quantised energy levels and that the energy gaps between them can be measured directly using optical spectroscopy. If the energy of the lowest state is known then this allows absolute values for the energies of the systems to be ascertained. In all cases but one, the energy due to vibration in a molecule (see below), the lowest energy is defined to be zero. In addition, assuming the Boltzmann distribution for the populations of the energy levels in systems at thermal equilibrium with their surroundings allowed us to calculate the physical properties of some two-level systems. We could take the agreement between results so calculated and experiment as evidence that the distribution is correct. But the theoretical derivation of the distribution gives insight into the conditions under which it is valid, and we return to this below. Meanwhile we continue to demonstrate that the simple approach we have used with two-level systems can be generalised to ones containing any numbers of levels.

2.1   MOLECULAR PARTITION FUNCTION

The energy of any quantised system can be obtained, as stated above, by summing over the populations of the individual states multiplied by the energies of those states:

$$E = \sum_i n_i \varepsilon_i \tag{2.1}$$

This is, though, an impractical formula for estimating E for systems, including molecules, in which the number of energy levels may be very large indeed. First, it involves a summation over terms in each of these and, second, we appear to need to know the energies of all of states. However, we have already seen this is not necessarily the case. Since the populations (n_i) depend on $e^{-\varepsilon_i/kT}$, any terms in the expansion of the sum for which $\varepsilon_i \gg kT$ have near-zero values of n_i can be neglected. This

limits the number that needs be considered and has far-reaching conse-
quences.

A common approximation is to assume that the various modes of
energy that a molecule may possess are independent, so that for each level

$$\varepsilon_{total} = \varepsilon_{electronic} + \varepsilon_{vibrational} + \varepsilon_{rotational} + \varepsilon_{translational} \qquad (2.2)$$

This is the same assumption made in interpreting molecular spectra. It is
a very good approximation but it *is* one and there are many cases in
which it can be seen to be (slightly) inaccurate. A simple example is that a
change in the vibrational energy of an anharmonic oscillator (*e.g.* a real
diatomic molecule) involves a jump to a higher vibrational level in which
the average internuclear distance differs from that in the lowest one, and
therefore causes a change in the moment of inertia and the rotational
energy of the molecule, too. However, the approximation is sufficiently
good for us to be able to understand the properties of molecules and to
calculate their thermodynamic properties, often within measurement
accuracy. If we require their exact values we have no alternative to
measuring the energy levels experimentally and entering them into the
un-approximated energy equation, without assuming that the individual
contributions are independent.

Summing over the known energy levels is the most convenient way of
obtaining the energies of two- and three-level systems. For multi-level
systems, however, this approach is impractical and we turn to a second
method that involves investigating the denominator of the Boltzmann
distribution, to which we give a special name, the *molecular partition
function, q*:

$$q = \sum_i g_i e^{-\varepsilon_i/kT} \qquad (2.3)$$

From Chapter 1, it is apparent that as $T \to 0\,K$, $\varepsilon/kT \to \infty$ and $e^{-\infty} = 0$,
so that at $0\,K$ the expansion contains only one term and $q = g_0$, the
degeneracy of the ground state, a small number, often 1. However, as
$T \to \infty$, the exponentials in all the terms in the expansion tend to 1 and

$$q \to \sum_i g_i \qquad (2.4)$$

This is the total number of states in the system. Thus q has the physical
interpretation that it represents the number of states that are occupied at
a given temperature. It tends to infinity at very high temperatures (strictly
when $\varepsilon_i/kT \to 0$ for all i).

Why q is a valuable concept follows from doing a mathematical manipulation on it. We differentiate it with respect to T:

$$\frac{\mathrm{d}q}{\mathrm{d}T} = \sum_i \frac{\mathrm{d}(g_i \mathrm{e}^{-\varepsilon_i/kT})}{\mathrm{d}T} = \sum_i g_i \frac{\varepsilon_i}{kT^2} \mathrm{e}^{-\varepsilon_i/kT} \tag{2.5}$$

But for 1 mole of gas,

$$E = \sum_i n_i \varepsilon_i = \sum_i \frac{N_A g_i \mathrm{e}^{-\varepsilon_i/kT}}{q} \varepsilon_i \tag{2.6}$$

where we have written the Boltzmann distribution (in terms of q) to obtain n_i.

Combining the two equations then gives

$$E = \frac{RT^2}{q} \frac{\mathrm{d}q}{\mathrm{d}T} = RT^2 \frac{\mathrm{d}\ln q}{\mathrm{d}T} \tag{2.7}$$

This shows that E can be obtained directly from q. There remains one further step. Above, we defined the lowest quantum state to have an energy of zero. This is fine for the electronic energy of the system, for we remind ourselves that in experiment we often can only measure change and an arbitrary zero is a convenience. But one of the internal energies of molecules, the vibrational energy, has a non-zero value with respect to this electronic energy zero even at 0 K, and vibration is said to possess 'zero-point energy'. Quantum mechanics applied to a simple harmonic oscillator shows that its value is $\frac{1}{2}hv$ per molecule, where v is the vibrational frequency. We must add any contribution due to zero-point energy to that we calculated above to get the total energy, termed the 'Internal Energy' of the system. For 1 mole of material,

$$U(T) = U(0) + E = U(0) + RT^2\left(\frac{\delta \ln q}{\delta T}\right)_V \tag{2.8}$$

where $U(0) = \frac{1}{2} N_A hn$. Here we have added the subscript V to remind ourselves (see the comments on the Boltzmann equation in Chapter 1) that this is a result for a constant volume system and we use partial differentials because all variables other than T are held constant. C_V is properly defined as

$$C_V = \left(\frac{\partial U}{\partial T}\right)_V \tag{2.9}$$

but since $U(0)$ does not depend on temperature the results obtained for this quantity in Chapter 1 remain correct.

$U(T)$ may therefore be calculated if the energy levels of the system are known and it appears that we have reduced a very long series to an algebraic expression. However, this is an illusion since the partition function itself is a similarly long series. We shall see below, though, that once we invoke the separability of the different contributions to the energy, and use standard models to calculate the vibrational, rotational and translational energies of molecules, we can indeed reduce the calculation to algebraic form.

2.2 BOLTZMANN DISTRIBUTION

We now consider the assumptions that underlie the Boltzmann distribution. We have already stated that it applies to a system consisting of a constant number of independent non-interacting molecules. Its basis lies in the statistics of large numbers and on the assumption that all the different possibilities for how molecules may be distributed amongst their quantum states to give the same overall energy are equally likely. These distributions are characterised by *configurations*, themselves defined by the populations of the various levels (n_i, where $i = 0, 1, 2$ *etc.*), and some are more likely than others. A simple example was seen in Chapter 1 where in a two-level system occupied by two particles one configuration, that with one particle in each level, could be attained in two ways whereas the configurations with two particles in either the upper or the lower level could be obtained in just one. In general a configuration is characterised by its *statistical weight*, W defined by

$$W = \frac{N!}{n_0! n_1! n_2! \ldots} \tag{2.10}$$

which is the number of different ways in which it can be obtained. This follows because the permutation of particles within a given energy level does not produce a new distribution. In this case $N!$ is the number of permutations of N whereas the number of permutations of the numbers in each level is given by the product of permutations of the number in each level. As the total number of molecules increases the probability that some configurations become much more likely than others rises rapidly (see Problem 2.1). For the very large numbers of molecules that we usually deal with in chemistry (remember that 1 mole contains 6.022×10^{23} molecules) *just one* configuration becomes so much more likely than the others that at any time the system exists only in this dominant configuration and only it determines the properties of the system. This seems strange, for we are unfamiliar with the mathematical properties of very large numbers in everyday life but it leads to the Boltzmann distribution, which can be verified experimentally and appears to be correct.

We can determine what this dominant configuration is by seeking the values of n_i that maximise W, although it proves mathematically simpler to maximise $\ln W$, given by

$$\ln W = \ln N! - \sum_i \ln n_i! \tag{2.11}$$

Using Stirling's approximation ($\ln x! \approx x \ln x - x$) and remembering that $\Sigma_i \, n_i = N$, this yields

$$\ln W = N \ln N - \sum_i n_i \ln n_i \tag{2.12}$$

But in this configuration the total energy of the system remains

$$E = \sum_i n_i \varepsilon_i \tag{2.13}$$

and the total number of molecules is still

$$N = \sum_i n_i \tag{2.14}$$

These two equations imply that the populations of the states are not independent of each other. For instance, if a molecule moves from the first energy level to the second, another must do the reverse to conserve the total energy.

A change in $\ln W$ with a change in n_i can be expressed as

$$\mathrm{d} \ln W = \sum_i \left(\frac{\partial \ln W}{\partial n_i} \right) \mathrm{d}n_i = 0 \tag{2.15}$$

when $\ln W$ is a maximum. This equation follows from the fact that an infinitesimal change in n_i to $(n_i + \mathrm{d}n_i)$ causes a change in $\ln W$ to $\ln W + \mathrm{d} \ln W$. The two constraints demand that when the populations n_i change $\Sigma_i \mathrm{d}n_i = 0$ and $\Sigma_i \varepsilon_i \mathrm{d}n_i = 0$. We have, therefore, to find the most probable value of $\ln W$ under the conditions that all three equations involving $\mathrm{d}n_i$ are satisfied. This is accomplished using an elegant general method due to Lagrange.

We add the latter equations to the first, introducing two constant multipliers, α and β, whose values have to be determined to make the approach valid:

$$\mathrm{d}\ln W = \sum_i \left(\frac{\partial \ln W}{\partial n_i}\right) \mathrm{d}n_i + \alpha \sum_i \mathrm{d}n_i - \beta \sum_i \varepsilon_i \, \mathrm{d}n_i = 0 \qquad (2.16)$$

Here the negative sign has been introduced to ensure that β turns out to be a positive quantity, purely for convenience. Since the molecules are independent so are all changes in their numbers (*i.e.* $\mathrm{d}n_i$) and it follows that for each value of i

$$\frac{\partial \ln W}{\partial n_i} + \alpha - \beta \varepsilon_i = 0 \qquad (2.17)$$

Differentiating the equation for $\ln W$,

$$\frac{\partial \ln W}{\partial n_i} = -\ln n_i - 1 \approx -\ln n_i \qquad (2.18)$$

since n_i is a very large number. When substituting this into Equation (2.17) our focus changes. Instead of determining the maximum value of $\ln W$, what we shall now obtain is an expression for the values of n_i that ensure that $\ln W$ is a maximum. Re-arranging the resulting equation gives

$$n_i = \mathrm{e}^{(\alpha - \beta \varepsilon_i)} = \mathrm{e}^{\alpha} \mathrm{e}^{-\beta \varepsilon_i} \qquad (2.19)$$

and

$$N = \sum_i n_i = \mathrm{e}^{\alpha} \sum_i \mathrm{e}^{-\beta \varepsilon_i} \qquad (2.20)$$

where the first term can be taken out of the sum because e^{α} contains no term in i. Manipulation of these two equations gives

$$n_i = N \frac{\mathrm{e}^{-\beta \varepsilon_i}}{\sum_i \mathrm{e}^{-\beta \varepsilon_i}} \qquad (2.21)$$

This is the Boltzmann distribution if we put $\beta = 1/kT$; this will be justified later (Section 2.5.4.1). Notably, it is expressed here in terms of sums over quantum states, *i.e.* in terms of individual ε_i values. But each state may be degenerate and consist of many levels of equal energy; to account for this we have expressed the distribution throughout this book in terms of levels rather than states, *i.e.* we have included the degeneracies, g_i, in our equations.

We have assumed in the derivation that the molecules are independent and non-interacting. Now we generalise our treatment to the case where they might interact.

2.3   CANONICAL PARTITION FUNCTION

In many real systems molecules interact, for example through inter-molecular forces, and/or change their positions in a crystal lattice to become distinguishable from one another. The interactions allow them collectively to explore all the possible energy states available to the molecules individually, with the total energy of the system conserved if it is isolated from its surroundings. Our focus changes from that of the Boltzmann distribution in that we are concerned now with the distribution of energy between the whole series of possible energies for the *complete* system containing all the molecules, rather than how the molecules are distributed between their individual quantum states. This is because at any one instant the collection can adopt any one of a range of energies, depending on the precise distribution of molecules between their quantum states. Sometimes the total energy will exceed the average and sometimes will be less than it. But over time the energy, which we wish to calculate, is constant. The situation is one of constant dynamic change with molecules jumping between states due to their interactions (*e.g.* in collisions) and we must consider all the possible distributions that are explored, and take their average over time to obtain the mean properties of the system. There is no known way of doing this and so we resort to an elegant alternative. We simply take all the different possible alternative distributions that would occur over all time and pretend that they all exist at the same instant, and then take their average. They are said to constitute an *ensemble*. An experiment described in Chapter 1 demonstrated that, provided that thermal equilibrium is maintained, the spectrum of a single molecule observed over time is identical to that exhibited by a large collection of molecules instantly, and provides direct evidence that the approach might be sound. It turns out that, as with the Boltzmann distribution itself, with a very large number of alternative distributions one becomes so much more likely than the others that only it need be included in the calculations. That is, the averaging includes only it.

So much for the basic idea, but now we should be careful and strict. We define a *Canonical Ensemble* as one composed of a very large number of identical and imagined systems (*i.e.* collections of molecules) of fixed N, V and T, all of which are in good thermal contact with each other so that they can exchange energy. Each possesses its own energy E_i but the whole ensemble is thermally isolated from its surroundings so that its total energy is constant. We let the total number of systems be N^* and the total energy of the whole ensemble be E^*. The ensemble has a total volume of N^*V, which is fixed. The calculation is mathematically very similar to that of the Boltzmann distribution, and we shall not repeat it, but here it tells us how to achieve the dominant configuration of the various systems in the ensemble between their various possible energy states.

The *Canonical distribution* is

$$\frac{n_i^*}{N^*} = \frac{e^{-E_i/kT}}{\sum_i e^{-E_i/kT}} \qquad (2.22)$$

where n_i^* is the number of systems of energy E_i and the average energy of a system is given by

$$E = \frac{E^*}{N^*} = \frac{\sum_i n_i^* E_i}{N^*} \qquad (2.23)$$

We define a canonical partition function as

$$Q = \sum_i e^{-E_i/kT} \qquad (2.24)$$

The Internal Energy of the system is given by

$$U(T) = U(0) + kT^2\left(\frac{\partial \ln Q}{\partial T}\right)_V \qquad (2.25)$$

This is obtained in an analogous fashion to that used above to obtain an expression containing q but this equation is perfectly general and not restricted to independent non-interacting molecules.

Q is defined in terms of the total energies of the imaginary systems that comprise the canonical ensemble but each of these consists of component molecules whose energies are quantised, and this suggests that a relation exists between Q and q. If the molecules are independent then the energy of a system of N molecules is given by

$$E_i = (\varepsilon_1 + \varepsilon_2 + \ldots + \varepsilon_N)_i \qquad (2.26)$$

where $\varepsilon_1, \varepsilon_2$ *etc.* are the energies of molecules 1,2 *etc.* in their quantum states in the particular system, *i*. In consequence

$$Q = \sum_i e^{-E_i/kT} = \sum_i e^{-(\varepsilon_1 + \varepsilon_2 + \ldots \varepsilon_N)_i/kT} = \sum_i e^{-(\varepsilon_1)_i/kT} \sum_i e^{-(\varepsilon_2)_i/kT} \cdots \sum_i e^{-(\varepsilon_N)_i kT}$$

(2.27)

Each of the terms in this product involves one specific molecule, but we have seen above that over time a molecule explores all its possible quantum states so that the value of each term is equal to the value obtained by considering the whole assembly of molecules at a specific instant. This is the molecular partition function, q. Hence,

$$Q = q^N \qquad (2.28)$$

An alternative way of deriving this result with $N = N_A$ is to realise that, for a system of independent molecules, the two expressions we have obtained for the Internal Energy must be equivalent:

$$U(0) + kT^2 \left(\frac{\partial \ln Q}{\partial T}\right)_V = U(0) + RT^2 \left(\frac{\partial \ln q}{\partial T}\right) \qquad (2.29)$$

and $R = kN_A$.

This relation between Q and q is valid provided that the molecules are not only independent but distinguishable. That is, they must be different chemical species in the gas phase, or attached to different lattice points in a crystal. If we can distinguish which molecule is which we know which is in a particular energy state inside a particular system within the ensemble. If the molecules cannot be distinguished we do not. In this case a particular energy state, E_i, can be obtained in several different ways. For example, in our two-level system containing two molecules discussed in Chapter 1, the energy ε could be obtained in two ways, with either molecule in the upper energy state and the other in the lower one. There are, therefore, two ways of obtaining this energy value in this very small ensemble, whereas we are concerned here only with how many different total energy values exist. If we include both we consequently overestimate their possible number.

To correct for over-counting at all but exceptionally low temperatures the expression that must be used when considering indistinguishable particles is

$$Q = \frac{q^N}{N!} \qquad (2.30)$$

This is not easy to derive but some rationale for it is as follows: at high temperatures each of the N molecules can enter any one of a large number of quantum levels, and the total number of such levels in the system therefore greatly exceeds the number of molecules. The number of ways of forming a system of N molecules is then the number in which they can enter their individual quantum states, *i.e.* the number of ways of assigning N particles to N quantum states, all of which are accessible at high temperature. With each chosen at random this is $N!$

2.4 SUMMARY OF PARTITION FUNCTIONS

The Internal Energy of a system (and we shall see all other physical properties) can be calculated using the canonical partition function, Q. This is whether or not the molecules (or atoms) interact. In the particular, but common, case that they do not interact but are independent there is a direct relation between Q and q, the molecular partition function. This relationship differs according to whether the molecules are distinguishable or indistinguishable, but in either case we can calculate the physical properties from a knowledge of the quantised energy levels of the molecules. In this way partition functions provide a crucial and powerful link between apparently disparate subjects within physical chemistry, and it implies that we can understand the bulk properties of materials in terms of the energy levels of the atoms or molecules they contain.

We must remember, however, that the distributions and partition functions resulted from the statistical behaviour of very large numbers of molecules, or systems inside the canonical ensemble. Most experiments performed before the last decade satisfied this criterion, but newer ones involve small numbers of molecules, even at times a single one, and we should not assume that the distributions still apply. We mentioned above that the spectrum of a single molecule observed over a long period of time during which it is in good thermal contact with its surroundings, is identical to that of an ensemble of molecules, since after each transition it gives energy to its surroundings and drops back down to its original level. However, if the thermal contact is broken and the molecule cannot change its quantum state in this way then its spectrum is a single line corresponding to a single transition between two quantum states.

2.5 EVALUATION OF MOLECULAR PARTITION FUNCTIONS

Calculation of the physical properties of systems requires us to calculate Q, which can be obtained from q. To calculate the latter we need values of the actual energy levels. For the most accurate calculations this cannot be avoided and in what follows we should remember that this route is always open to us. However, it would be convenient for investigating how the properties depend on the nature of the specific type of molecule concerned if we could replace the essentially infinite sum in the definition by an algebraic expression that had such a dependence. This is accomplished through a series of approximations. Throughout what follows we shall test how good or bad these approximations, which are very commonly made, are.

The first is the separability of molecular energies into independent contributions from electronic, vibrational, rotational and translational terms, as discussed above. This allows us to re-write q:

$$q = \sum_i g_i e^{-\varepsilon_i/kT} = \sum_i g_i e^{-(\varepsilon_{el} + \varepsilon_{vib} + \varepsilon_{rot} + \varepsilon_{trans})_i/kT} \qquad (2.31)$$

Associated with each is a degeneracy, and the total degeneracy, is the product of them all so that we may factorise this expression:

$$q = \sum_{el,i} g_{el,i} e^{-\varepsilon_{el,i}/kT} \sum_{vib,i} g_{vib,i} e^{-\varepsilon_{vib,i}/kT} \sum_{rot,i} g_{rot,i} e^{-\varepsilon_{rot,i}/kT} \sum_{trans,i} g_{trans,i} e^{-\varepsilon_{trans,i}/kT} \qquad (2.32)$$

Each term has the same form as q itself and we write, with obvious notation,

$$q = q_{el} q_{vib} q_{rot} q_{trans} \qquad (2.33)$$

That is, the total partition function can be written as a product of individual partition functions that originate in each type of energy that the molecule possesses. This is a general result that appears whenever the total energy can be written as the sum of independent contributions. It implies that we can evaluate each independently and we find below that doing this, with further assumptions, turns the infinite series in q into an easily evaluated algebraic function. The principles involved are most readily appreciated by considering the smaller contributions first.

2.5.1   Electronic Partition Function

To calculate the electronic partition function we use the energies of the atomic or molecular levels measured using electronic spectroscopy in the expansion

$$q_{el} = \sum_{el} g_{el} e^{-\varepsilon_{el}/kT} = g_0 e^{-0/kT} + g_1 e^{-\varepsilon_1/kT} + g_2 e^{-\varepsilon_2/kT} + \cdots \qquad (2.34)$$

where 0 is the (defined) energy of the lowest electronic level and ε_1, ε_2 *etc.* are the energies of the higher levels. Provided that the levels are not split by spin–orbit coupling, then the energies are those of the atomic or molecular orbitals. These are well separated in energy from the ground state so that all the higher energies are $\gg kT$ at room temperature with the result that they contribute negligibly to the sum and can be ignored. This leaves only the first term so that

$$q_{el} = g_0 \qquad (2.35)$$

The electronic partition function is therefore often a small number equal to the degeneracy of the lowest state. It is independent of temperature, implying that the electronic heat capacity near room temperature is zero. This changes only in the presence of spin–orbit coupling, in for example the halogen atoms and the NO molecule, which creates a low-lying higher level so that two terms have to be included in the expansion at room temperature.

2.5.2 Vibrational Partition Function

We could adopt exactly the same approach of expanding the partition function and see which terms were significant in evaluating q_{vib}. Indeed, for most diatomic molecules at room temperature only the first, singly-degenerate term is significant. However, vibrational levels are much closer together than electronic ones, and further terms may contribute as the temperature is increased. The temperature at which this happens depends on the precise nature of the molecule involved; we can understand why by using a model for a molecular vibrator, which we treat as a Simple Harmonic Oscillator (SHO), to obtain an algebraic expression for q_{vib}. This is the second approximation in evaluating q.

From quantum mechanics, the energy of a SHO is given by

$$\varepsilon_{vib} = (v + \tfrac{1}{2})h\nu \qquad (2.36)$$

where v is the vibrational quantum number ($= 0, 1, 2 \ldots$) and ν is the vibrational frequency. The degeneracy of each vibrational state is 1. This implies a set of equally spaced levels separated by $h\nu$ (Figure 2.1). The frequency is given by

$$\nu = \frac{1}{2\pi}\sqrt{\frac{k}{\mu}} \qquad (2.37)$$

where k is the bond force constant and μ is the reduced mass (Section 1.3).

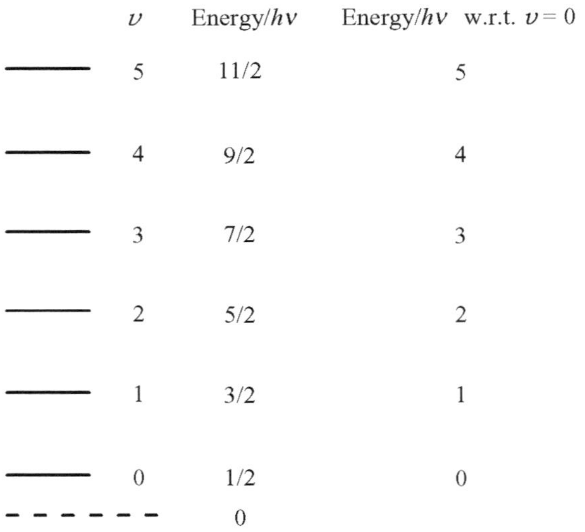

v	Energy/hv	Energy/hv w.r.t. $v=0$
5	11/2	5
4	9/2	4
3	7/2	3
2	5/2	2
1	3/2	1
0	1/2	0
	0	

Figure 2.1 *The first five energy levels of a Simple Harmonic Oscillator. The levels are equally spaced with a separation of* hv *but the lowest has an energy of* $\frac{1}{2}hv$ *with respect to the lowest electronic level (shown as a broken line). This is the zero-point energy and a molecule cannot possess a lower energy than this. When just the vibrational properties of molecules are considered it is convenient to define the energy of the lowest vibrational level (* v = 0 *) to be zero, as shown on the right.*

This is where the properties of the molecule enter the calculation. Whether this is a good model can only be checked experimentally, and in precise spectroscopic measurements we find that a correction to the energy is needed to account for anharmonicity; that is the vibration is not accurately simple harmonic. However, this correction is small and we can safely neglect it when trying to understand the implications to physical properties.

The energy equation shows that when the molecule is in its lowest vibrational level ($v = 0$) it still has a 'zero-point energy' of $\frac{1}{2}hv$. Following our convention, however, if we are considering changes involving vibrational energy alone we define *this* level to have zero energy and relative to it the SHO levels have energies of

$$\varepsilon_{\text{vib}} = vhv \tag{2.38}$$

so that

$$q_{\text{vib}}^{(0)} = \sum_v e^{-vhv/kT} = e^0(=1) + e^{-hv/kT} + e^{-2hv/kT} + \ldots \tag{2.39}$$

where we have introduced the superscript (0) to denote that this is with respect to the zero-eth vibrational level having zero energy. This still contains an infinite series but it is a geometric progression with a constant coefficient $e^{-hv/kT}$ and its sum is a standard algebraic result:

$$q_{\text{vib}}^{(0)} = \frac{1}{1 - e^{-hv/kT}} \tag{2.40}$$

This is simple to evaluate at any temperature if v is measured. Experiment shows that only the ground vibrational state is occupied in most diatomic molecules at room temperature, implying that $hv \gg kT$ and at room temperature for many diatomic molecules $q_{\text{vib}} \approx 1$.

What we have done here may be regarded with suspicion, for the vibrating molecule is in an electronic state whose lowest energy we have already defined as 0 (Figure 2.1). We have therefore defined two quite separate energy zeroes in the molecule, and in fact the lowest vibrational energy is $\frac{1}{2}hv$ in energy above the zero of the electronic level. Where properties such as the Internal Energy are concerned this is simplest dealt with, as above, by adding the temperature-independent zero-point energy to the energy equation. But the full treatment is obtained simply by factoring the basic definition of the partition function:

$$q_{\text{vib}} = \sum_{v} e^{-(v + 1/2)hv/kT} = e^{-1/2hv/kT} \sum_{v} e^{-vhv/kT} = e^{-1/2hv/kT} q_{\text{vib}}^{(0)} \tag{2.41}$$

If this is inserted into the equation for the energy, E (above), then the result about adding the zero-point energy falls out and we do not need to introduce this separately.

Whereas a diatomic molecule has only one vibrational frequency,

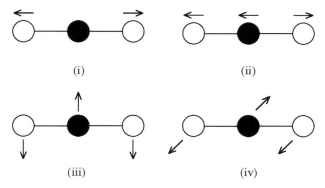

(i)　　　　　　　　　(ii)

(iii)　　　　　　　　　(iv)

Figure 2.2 *The four independent normal modes of vibration of a linear tri-atomic molecule such as CO_2. (i) Symmetric stretch, (ii) asymmetric stretch, (iii) and (iv) two degenerate bending vibrations; (iii) occurs in the plane of the paper and (iv) out of the plane of the paper.*

polyatomic molecules undergo a number of independent vibrations. For example, in a linear tri-atomic molecule such as CO_2 there are four (Figure 2.2). One is a symmetric stretch (ss), with the O atoms breathing symmetrically on either side of the C atom, one is an asymmetric stretch (as), with two atoms moving in one direction and the third in the opposite direction, and there are two perpendicular and independent bending (b) vibrations, which are degenerate. Each type of vibration occurs with its own frequency and energy, and the total energy is their sum. It follows that, by analogy with our previous separation of energy terms, the total vibrational partition function of the molecule is given by the product of partition functions for each:

$$q_{vib} = q_{ss}q_{as}q_{b}q_{b} = \left(\frac{e^{-1/2h\nu(ss)/kT}}{1 - e^{-h\nu(ss)/kT}}\right)\left(\frac{e^{-1/2h\nu(as)/kT}}{1 - e^{-h\nu(as)/kT}}\right)\left(\frac{e^{-1/2h\nu(b)/kT}}{1 - e^{-h\nu(b)/kT}}\right)^2 \quad (2.42)$$

This is again perfectly straightforward to evaluate if the three different vibration frequencies are measured.

2.5.2.1   Vibrational Heat Capacity of a Diatomic Gas.   We are now in a position to test the theory. We saw in Chapter 1 that vibration made an increasing contribution to the heat capacity of a diatomic gas as the temperature was raised. The vibrational contribution tends to an asymptotic value of $2 \times \frac{1}{2}R$ J mol^{-1} K^{-1} at high temperature, with H_2 and Cl_2 approaching this at different rates. Is this predicted by our theory?

The Internal Energy is given by

$$U(T) = U(0) + kT^2\left(\frac{\partial \ln Q}{\partial T}\right)_V \quad (2.43)$$

so that

$$U(T) = U(0) + RT^2\left(\frac{\partial \ln q_{vib}}{\partial T}\right)_V = U(0) + RT^2\frac{\partial}{\partial T}\left(\ln\frac{1}{1 - e^{-h\nu/kT}}\right) = \frac{RT\left(\frac{h\nu}{kT}\right)}{e^{h\nu/kT} - 1} \quad (2.44)$$

where we have used $Q = q^N$ since we can identify the diatomic molecule that possesses vibrational energy, and have multiplied top and bottom of the equation by a positive exponential term. Noting that by cancellation the numerator is independent of T, whilst $U(0)$ is not temperature-dependent, further differentiation to obtain C_V is straightforward:

$$C_V = \left(\frac{\partial U}{\partial T}\right)_V = \frac{R\left(\frac{hv}{kT}\right)^2 e^{hv/kT}}{(e^{hv/kT} - 1)^2} \qquad (2.45)$$

At low temperatures, $T \to 0$ and $(hv/kT) \to \infty$, as does the exponential, but faster. So $C_V \to 0$, as observed. At high temperatures $(hv/kT) \ll 1$ and we can expand the exponentials ($e^x = 1 + x/1! + x^2/2! + ...$) and truncate the expansion after the first term:

$$C_V \approx \frac{R(hv/kT)^2(1 + hv/kT)}{(hv/kT)^2} = R + R(hv/kT) \approx R \qquad (2.46)$$

This is exactly the value predicted by the Equipartition Theorem, and is in accord with experiment. It rationalises why the system behaves classically at high temperatures.

The full temperature dependence, shown in Figure 2.3, and it reproduces the behaviour observed experimentally.

The treatment also shows why H_2 and Cl_2 reach this asymptotic limit at different temperatures. Dihydrogen is composed of lighter atoms than Cl_2 and so, from the SHO equations, has the higher vibrational frequency. Insertion of the frequencies of the two molecules into the heat capacity equation yields a quantitative fit to each within experimental error. This is a remarkable achievement for there are no variables in the equation other than T. The basic reason for the molecules attaining the limiting C_V at different temperatures is that the gap between the vibra-

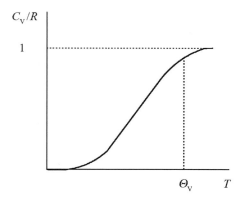

Figure 2.3 *Theoretically predicted variation of the vibrational heat capacity with temperature. It reproduces the experimental observations on a given molecule when its parameters are included in the calculation. In contrast to what appears to be the case in the schematic plot shown in Figure 1.4 it is apparent that by the time* T = Θ_V *(p. 52) some 90% of the maximum vibrational contribution is made.*

tional levels in H_2 is greater than that in Cl_2 and so T has to be increased by more to satisfy the condition that $kT \approx$ the energy gap before the vibrational energy of the molecules can change.

2.5.3 Rotational Partition Function

With rotation it is usually no longer practical to evaluate the partition function simply by expanding its definition and evaluating each term. As shown in Chapter 1, rotation makes its full contribution to the heat capacities of diatomic gases at room temperature, which means that the energy of the system must have attained its asymptotic, high temperature, limit. In turn this implies that $kT \gg$ the energy gaps between the various rotational levels associated with the lowest vibrational level of the lowest electronic state. Many rotational levels are therefore occupied and summation to obtain the partition function may be over a large number of them, although we shall see below that we have to be careful in assuming this.

Our approach is once more to take a specific model, for a rotating diatomic molecule the rigid rotor. This is another approximation. Quantum mechanics tells us that its energies are

$$\varepsilon_{rot} = hcBJ(J + 1) \tag{2.47}$$

where J is the rotational quantum number $(= 0, 1, 2 \ldots)$ and B is the rotational constant given by

$$B = \frac{h}{8\pi^2 Ic} \, \text{cm}^{-1} \tag{2.48}$$

and the moment of inertia for a diatomic molecule can be expressed as

$$I = \mu \langle r^2 \rangle \tag{2.49}$$

This reminds us that the square of the internuclear distance should be averaged over the vibration of the molecule, the molecule making about 1000 vibrations in one rotation. As with vibrational energy, accurate spectroscopic observations tell us that we should correct the model, in this case to account for centrifugal stretching of the bond at high rotational energies (high J numbers), but the correction is again small and has little effect on physical properties. It can conveniently be neglected in our calculations (see Problem 2.4).

In contrast to the vibrational energy levels of a SHO, rotational levels are not equally spaced but rather occur at 0, $2B$, $6B$, $12B$ *etc.* (Figure 2.4).

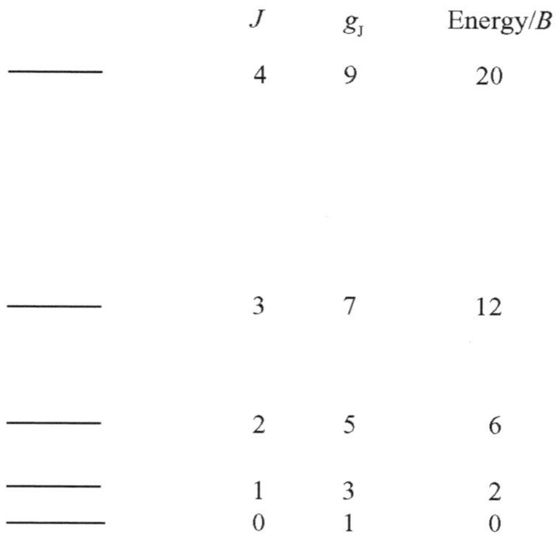

J	g_J	Energy/B
4	9	20
3	7	12
2	5	6
1	3	2
0	1	0

Figure 2.4 *Associated with each vibrational energy level there is a family of rotational levels, the first five of which are shown in the rigid rotor model. In contrast to the vibrational levels in Figure 2.1 these levels are not equally spaced in energy but the gaps between the levels remain much smaller than those between vibrational levels. Each rotational level is (2J + 1)-fold degenerate.*

However, for diatomic molecules $B \ll kT$ at room temperature (typically 1–10 cm^{-1} compared with 207 cm^{-1}) and so, as stated above, many levels may be occupied. The temperature at which this inequality is satisfied as it is increased from 0 K again depends on the specific molecule concerned, through the inverse dependence of B on the reduced mass, μ. As with vibration, this happens at a lower temperature for dichlorine than for dihydrogen.

In further contrast to the vibrational energy levels of diatomic molecules the rotational levels have degeneracies, given by $g_{rot} = (2J + 1)$, so that

$$q_{rot} = \sum_J (2J + 1) \, e^{-hcBJ(J + 1)/kT} \tag{2.50}$$

These degeneracies have a profound effect on the populations of the energy states. This can be seen as follows: at a given temperature q_{rot} is a constant and so, through the Boltzmann distribution, the population of state, n_J, is

$$n_J = \frac{N(2J + 1)e^{-hcBJ(J + 1)/kT}}{\text{const}} \tag{2.51}$$

Whereas the exponential term decreases with increasing J, the degener-

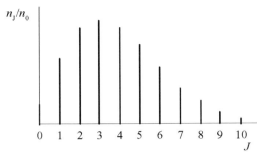

Figure 2.5 *The population* n_J *of the rotational levels relative to the population of the lowest level associated with the lowest vibrational state of a diatomic molecule plotted against the rotational quantum number at room temperature. The graph is drawn for HBr at 298 K. If all the levels were singly degenerate this graph would simply show an exponential fall in population as* J *is increased, but the (2J + 1) degeneracy factor of the rigid rotor multiplies this by a linearly increasing function in* J. *The result is that the maximum population is not found in the lowest energy level but rather in a higher one, whose actual value depends on the precise molecule concerned. But the distribution is broad, implying substantial population in a number of rotational levels. The effect of this population distribution is seen in the vibration–rotation spectrum, Figure 2.6, although see the text.*

acy increases linearly with it and the population depends on the product of the two and maximises at a *J* value not equal to zero (Figure 2.5). That is, the state with the highest population is not the lowest one; this leads to the effect of degeneracy on populations that we saw in Chapter 1. This can be verified directly in the infrared, vibration–rotation, spectrum of a sample of molecules (Section 4.4). At room temperature this originates in transitions between the populated rotational states of the lowest vibrational level and the unpopulated rotational states of the first higher vibrational level, so that only absorptive transitions occur. The intensities of the lines therefore largely reflect the populations of the rotational levels of the ground vibrational state. This would be exactly so if each of the transitions allowed by the selection rules occurred with equal probability, but they do not. However, the populations of the levels remain a major factor. The spectrum (Figure 2.6) consists of two branches according to whether the *J* quantum number changes by +1 or −1, and each branch exhibits a maximum in its intensity, corresponding to involvement of the most highly populated state.

Since the rotational levels are comparatively close together we attempt a further approximation. We assume that the energies are essentially continuous, *i.e.* there is an infinity of levels, and this allows the infinite sum in the partition function to be replaced by an integral which when evaluated gives

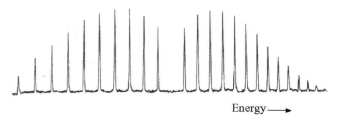

Energy⟶

Figure 2.6 *Vibration–rotation spectrum of HBr at room temperature. The transitions occur in two bands about the centre with the energy increasing from left to right. The lower energy band (the 'P' branch) arises from transitions between levels of quantum number J in the lower vibrational level (v = 0) to ones of J–1 in the upper level (v = 1), whilst the higher energy one (the 'R' branch) arises in transitions that increase this quantum number by one. Relative to the centre the transitions occur from the J = 0,1,2,3 . . . levels and the intensity maximises around J = 3–4 in both bands. This largely reflects the populations of the rotational levels in the vibrational ground state (Figure 2.5), that is their Boltzmann distribution. This is not exactly so since in detail the different transitions occur with slightly different probabilities, but the populations remain the dominant factor.*

$$q_{rot} \approx \frac{kT}{hcB} \tag{2.52}$$

and again a simple algebraic expression has been obtained that allows us to appreciate how the partition function varies between different molecules, which have different values of B.

The equation requires an amendment, the introduction of a *symmetry number*, σ:

$$q_{rot} = \frac{kT}{\sigma hcB} \tag{2.53}$$

This represents the number of times that a molecule acquires an indistinguishable configuration as it is rotated through 360°. For a heteronuclear diatomic (*e.g.* HCl) this is 1, but for a homonuclear diatomic (*e.g.* $^{16}O_2$) it is 2 – the molecule looks exactly the same after rotation through 180° as it does after rotation through 360°. The origins of this are fundamental and lie in the Pauli Principle, which requires the wave functions of molecules to obey strict symmetry rules. Its effect is to restrict a symmetric molecule such as $^{16}O_2$ to only half of the rotational energy levels available to a heteronuclear diatomic molecule (and ^{16}O–^{18}O is heteronuclear, for example). This has an obvious effect on q_{rot}, which is now summed over half as many states. Detailed application of the Pauli principle requires careful consideration of the nuclear spins of the component identical nuclei (Section 4.5).

So how good *is* the assumption that an infinity of rotational levels is occupied even at room temperature? Unfortunately, not as good as any of our previous approximations! Taking the measured B of HF, for example, yields a q_{rot} of about 10 at room temperature, indicating only this number of levels is occupied. Its true value, obtained by putting the measured energies of the levels into the primary definition of the partition function, differs by a few percent from this. For heavier molecules the approximation is much better since B varies as the inverse of the reduced mass, and q_{rot} may be several hundred, this many levels being able to be populated since they are now much closer in energy. However, the approximation allows us to understand how the partition functions and the physical properties of molecules depend on their rotational constants, and therefore on their atomic constitutions and bond lengths.

It is sometimes convenient to express q_{rot} in terms of contributions from the different degrees of rotational freedom in molecules. As discussed in Chapter 1, the total energy can be written as the sum of independent contributions for rotation about three orthogonal Cartesian axes, x, y and z. It immediately follows that the partition function can be written as a product (see the separation of energy contributions, above):

$$q_{rot} = q_{rot,x} q_{rot,y} q_{rot,z} \qquad (2.54)$$

A diatomic molecule (or any linear one) has two equal and non-zero contributions to its energy from rotation about the x- and y-axes, but a zero contribution for rotation about the unique axis (Section 1.2). This implies, from the definition of the partition function, that $q_{rot,z} = 1$. We deduce from our expression for the partition function for rotation obtained above that

$$q_{rot,x} = q_{rot,y} = \left(\frac{kT}{\sigma h c B_x} \right)^{1/2} \qquad (2.55)$$

where we have used the fact that $B_x = B_y$.

This might be thought to allow a simple extension to polyatomic molecules with three different moments of inertia, simply by multiplying three square-root terms with three different Bs together. Unfortunately, this is not true because the equation we used for the rotational energy applied to a linear rigid molecule, which alone has a rotational energy that can be expressed in terms of a single quantum number. A symmetric top molecule such as CH_3Cl, for example, has rotational energy contributions from both the overall rotation of the whole molecule and from internal rotation about the unique axis within the molecule, both of which have associated quantum numbers. The calculation of the partition function is more complicated but yields

$$q_{rot} = \left(\frac{kT}{hc}\right)^{3/2} \left(\frac{\pi}{B_x B_y B_z}\right)^{1/2}$$

(2.56)

This differs only by $\pi^{1/2}$ from the value obtained by writing the product of three independent terms.

2.5.3.1 *Rotational Heat Capacity of a Diatomic Gas.*

Once more we test our theory by investigating whether it yields the Equipartition Principle value for the heat capacity in the high temperature limit. In this limit the use of integration rather than summation in calculating the partition function is justified and calculation of the rotational heat capacity is trivial.

Since rotational energy is localised in the molecules themselves and molecules with different rotational energies can be distinguished from one another, $Q = q^N$ and the internal energy is given by [since $U_{rot}(0) = 0$]

$$U_{rot} = RT^2 \frac{\partial \ln q}{\partial T} = RT^2 \frac{\partial}{\partial T}\left(\ln \frac{kT}{\sigma hcB}\right) = RT^2 \left[\frac{\partial}{\partial T}\ln\left(\frac{k}{\sigma hcB}\right) + \frac{\partial}{\partial T}\ln T\right] = RT$$

(2.57)

because only the second term is a function of T. It follows that

$$C_V = \left(\frac{\partial U_{rot}}{\partial T}\right)_V = R$$

(2.58)

as predicted by the Equipartition theorem, and in accordance with experiment. The result is independent of whether the diatomic is homonuclear or heteronuclear.

To obtain the heat capacity at lower temperatures we ultimately have little alternative but to expand the partition function from an experimental knowledge of the energies of the states, retaining all the terms that are significant at the temperature in question, which of course depends on the energy separation from the ground state as compared with kT. This approach contains no approximations and it accurately reproduces the experimental behaviour as the temperature is changed. However, considerable insight can be obtained by continuing to use our approximate method.

A convenient indication of where the rotational contribution becomes significant is provided by the 'rotational temperature, θ_{rot}', which is defined as hcB/k for a heteronuclear diatomic ($\sigma = 1$), and depends on the specific molecule concerned through its B value. With the exception of the unusually light HD, for which it is 65.7 K, this normally lies below 20 K. A 'vibrational temperature, θ_{vib}', defined for each molecule as $hc\bar{v}/k$, is the corresponding temperature for the vibrational contribution.

Partition functions are commonly written in terms of these temperatures. For rotation it becomes

$$q_{\text{rot}} = \sum_{J=0}^{\infty} (2J + 1)e^{-J(J+1)\frac{\theta_{\text{rot}}}{T}} \tag{2.59}$$

so that

$$U_{\text{rot}} = NkT^2\frac{\mathrm{d}\ln q_{\text{rot}}}{\mathrm{d}T} = \frac{Nk\theta_{\text{rot}}}{q_{\text{rot}}} \sum_{J=0}^{\infty} (2J + 1)J(J + 1)e^{-J(J+1)\frac{\theta_{\text{rot}}}{T}} \tag{2.60}$$

At low temperatures $T \leqslant \theta_{\text{rot}}$ and we may expand the sum:

$$U_{\text{rot}} \approx Nk\theta_{\text{rot}}\{0 + 6e^{-2\theta_{\text{rot}}/T} + \ldots\} \tag{2.61}$$

which tends to zero as $T \rightarrow 0$ K. Differentiating with respect to temperature gives

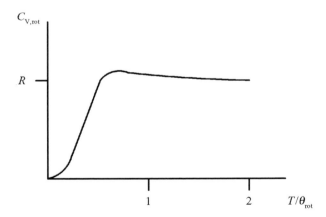

Figure 2.7 *Variation of the rotational heat capacity of a diatomic molecule with temperature. The value increases from zero in a way consistent with experiment and reaches an asymptotic value of R at high temperatures, in practice at about T $\geqslant$ 1.5 Θ_{rot}. Uniquely for rotation the equipartition value is exceeded at lower temperatures than this and the curve exhibits a maximum > R. This is the result of the degeneracy of the rotational levels increasing as the quantum number J increases. This effect is seen very clearly in the rotational contribution to the heat capacity of H$_2$ (see p. 87).*

$$C_{V,rot} = 12Nk\left(\frac{\theta_{rot}}{T}\right)^2\left\{e^{\frac{-2\theta_{rot}}{T}} + \ldots\right\} \tag{2.62}$$

which also tends to zero as $T \to 0\,K$.

The temperature dependence of $C_{V,rot}$ over the whole temperature range is obtained by expanding Equation (2.60) before differentiating. This causes the influence of the degeneracy factor to become apparent, as is shown in the normalised plot of Figure 2.7. Unique to rotation, as the temperature is raised, $C_{V,rot}$ increases from zero through a maximum before the high temperature limiting (equipartition) value is attained. The origin of the maximum lies in the $(2J + 1)$-fold degeneracy of the states that causes the maximum population not to be in the lowest energy level once θ_{rot} is approached as the temperature is increased. We saw this effect in two-level systems by showing that the population of the upper level could exceed that of the lower if the degeneracy of the state was greater. We return to this in Section 4.3.

2.5.3.2 *Rotational Partition Functions in the Liquid State.*

A partition function is defined only if the energy of a system is quantised. In the gas phase this raises no problems where rotation is concerned since each molecule rotates freely without interacting with those that surround it on the timescale of the rotation. Spectroscopy provides a direct measure of how fast this 'free' rotation is, and typically a single rotation occurs in about 10^{-10} s, a much shorter time interval than between molecular collisions. But in solution the molecules are much closer and they collide very frequently in their normal thermal motion. In a crude model we imagine that here the molecules almost touch each other and a collision occurs at every vibration of each molecule, about every 10^{-13} s. This is the actual timescale. So in attempting to make a single rotation a molecule is hit about 1000 times by each of its neighbours and gets knocked in random directions, preventing it from rotating freely. If free rotation cannot occur there are no quantised energy levels associated with it and the partition function cannot be written. Some molecules, in particular small spherical ones, can rotate freely in solution and then the function can be written. We conclude that equations involving the rotational partition function must be applied with caution in the liquid state. No such caution is necessary with either electronic or vibrational partition functions, whose natural timescales are very much faster than any other motions.

Molecules that do not rotate freely in solution nevertheless do rotate in very much slower rotational diffusion processes that are essential, for example, to the observation of nuclear magnetic resonance (NMR) spectra in solution, being involved in relaxation processes. These are processes that return a population difference that has been disturbed by a spectroscopic transition to its thermal equilibrium value by dumping energy to the surroundings.

2.5.4 Translational Partition Function

To write the translational partition function we need an expression for the quantised energy levels for translation. It is convenient to express the translational energy as the sum of individual terms from three independent degrees of freedom in orthogonal directions (Chapter 1):

$$\varepsilon_{trans} = \varepsilon_x + \varepsilon_y + \varepsilon_z \tag{2.63}$$

Writing the energy as a sum immediately implies that the partition function can be written as the product

$$q_{trans} = q_x q_y q_z \tag{2.64}$$

Now quantum mechanics comes to our aid once more as the 'particle in a box' calculation provides just the expression we need. It tells us that a particle of mass m confined in a one-dimensional box of length l_x has energy levels given by

$$\varepsilon_{1_x} = \frac{n_x^2 h^2}{8ml_x^2} \tag{2.65}$$

where n_x is a quantum number.
Consequently,

$$q_x = \sum_{n_x = 1}^{\infty} \exp\left\{\frac{n_x^2 h^2}{8ml_x^2 kT}\right\} \approx \int_0^{\infty} \exp\left\{\frac{n_x^2 h^2}{8ml_x^2}\frac{1}{kT}\right\} dn_x \tag{2.66}$$

Here we have replaced the infinite sum by the infinite integral to very high accuracy since $kT \gg$ the translational energy at all temperatures above a few fractions of K, and the energy levels are almost continuous. The initial sum is from the first occupied level ($n_x = 1$) but there are so many occupied levels that use of 0 in the integral introduces negligible error. The integral has a standard form:

$$\int_0^{\infty} e^{-\alpha n^2} dn = \frac{1}{2}\sqrt{\frac{\pi}{\alpha}} \tag{2.67}$$

and so

$$q_x = \frac{(2\pi mkT)^{1/2}}{h} l_x \qquad (2.68)$$

In three dimensions therefore

$$q_{\text{trans}} = \frac{(2\pi mkT)^{3/2}}{h^3} l_x l_y l_z = \frac{(2\pi mkT)^{3/2}}{h^3} V \qquad (2.69)$$

where V is the volume of the vessel containing the gas.

In a vessel of 100 cm^3 at room temperature and atmospheric pressure the translational partition function for the lightest of molecules, H$_2$, which has the smallest value, is consequently about 10^{24}. Even at 2 K it is of the order of 10^{21}. The approximation of replacing the infinite sum by the integral is an extremely accurate one except at temperatures very close indeed to 0 K. This is in contrast with the similar treatment applied to rotation, above, where the approximation produced significant errors.

2.5.4.1   The Translational Heat Capacity.   Translational motion ceases at 0 K and the internal energy at this temperature, $U(0)$, is therefore zero and so we have first to evaluate

$$U(T) = -kT^2 \left(\frac{\partial \ln Q_{\text{trans}}}{\partial T} \right)_V \qquad (2.70)$$

Due to the random motion, molecules are indistinguishable from one another and we take care to use the appropriate form for Q_{trans}.

$$\ln Q_{\text{trans}} = \ln \frac{q_{\text{trans}}^N}{N!} = N \ln q_{\text{trans}} - \ln N! \qquad (2.71)$$

Substituting for q_{trans} and expanding the logarithm gives

$$\ln Q_{\text{trans}} = \tfrac{3}{2}N\ln(2\pi mk) + \tfrac{3}{2}N\ln T + N\ln V - 3n\ln h - \ln N! \qquad (2.72)$$

Only the second of these terms is a function of T and so

$$\left(\frac{\partial \ln Q_{\text{trans}}}{\partial T} \right)_V = \frac{3N}{2T} \qquad (2.73)$$

and for 1 mole

$$U(T) = kT^2 \left(\frac{3N_A}{2T} \right) = \tfrac{3}{2}RT \text{ J mol}^{-1} \qquad (2.74)$$

so that

$$C_\mathrm{V} = \left(\frac{\partial U}{\partial T}\right)_\mathrm{V} = \tfrac{3}{2}R \mathrm{\ J\ K^{-1}\ mol^{-1}} \tag{2.75}$$

This is exactly the value predicted by the Equipartition theorem and is that observed experimentally over the whole temperature range above very close to 0 K.

This calculation is often used in the opposite sense. We could have retained the multiplier β introduced in deriving the Boltzmann expression in Section 2.2 and expressed the partition function in terms of it in the above derivation before equating the resulting internal energy to the $\tfrac{3}{2}RT$ value which is consistent with the heat capacity observed experimentally. This immediately leads to the result, quoted and used above, that $\beta = 1/kT$. In our derivation we used this relationship and obtained the result that agreed with experiment, so can conclude without further effort that it was correct.

2.6 OVERALL MOLECULAR PARTITION FUNCTION FOR A DIATOMIC MOLECULE

We simply substitute the expressions obtained for each individual factor into the equation

$$q_\mathrm{total} = q_\mathrm{el} q_\mathrm{vib} q_\mathrm{rot} q_\mathrm{trans} \tag{2.76}$$

to obtain

$$q_\mathrm{total} = g_0 \left(\frac{e^{-1/2 h\nu/kT}}{1 - e^{-h\nu/kT}}\right)\left(\frac{kT}{\sigma hcB}\right)\left(\frac{2\pi mkT}{h^2}\right)^{3/2} V \tag{2.77}$$

where we assume that the ground electronic state is not split by spin–orbit coupling. Note that we have been careful in writing the vibrational term to refer all quantised energies to the lowest energy of the ground electronic state, which is actually unattainable because of the zero-point energy term associated with vibrational motion. With only one type of molecule this does not affect a property such as the heat capacity. However, with more than one type of molecule, in calculating, for example equilibrium constants in reactions (see later), the zero-point point term should be included since it differs between the molecules involved. However, the effect of neglecting this is small, and the exponential term in the numerator is often omitted.

In deriving the expression for q_total several approximations have been made, but this expression is in algebraic form and can readily be enumerated if the spectroscopic constants of the molecule are known. Our approximate treatment based on specific models for molecular motion, and separability of the various contributions to the overall energy, provides real physical insight into what affects the behaviour and thermodynamic properties of the large collections of molecules that are usually involved in chemical and physical measurements.

We have seen that there can be significant, if small, limitations to the accuracy of this expression, particularly with the rotational term. We cannot expect it to yield exact answers but the approximations made are better for heavier molecules than light ones, and better at high temperatures than low. These are, however, limitations of the approximations and are not limitations to statistical thermodynamics itself. There is a true and absolute value to the partition function that can always be evaluated free of any assumptions or approximations from experimental measurements of the quantised energy levels that the molecules possess. In this way the properties of individual molecules and the thermodynamics of collections of them are intimately linked in a way that was not appreciated when thermodynamics was originally formulated.

APPENDIX 2.1   UNITS

All energies, including rotational and thermal energies, should be given in energy units. Yet we have seen in Section 1.7.3 that it is often convenient to express them as their equivalent in cm^{-1} by dividing by hc. When energies are directly ratio-ed, as in $e^{-\varepsilon/kT}$, this causes no confusion, but in this chapter we encounter a situation where the rotational constant B is *defined* to have units of cm^{-1} (Equation 2.48) and must be multiplied by hc to give true energy units. *i.e.* $\varepsilon_{rot} = hcBJ(J + 1)$. But if we insert this quantity into equations (for example Equation 2.50) containing kT the latter must now be in energy units. This causes derived equations such as Equation (2.53). ($q_{rot} = kT/\sigma hcB$) to be a little confusing. We have to remember that it is kT/hc that has units of cm^{-1}, and it is this quantity that has the value of 207.2 cm^{-1} at 298 K.

Another common source of difficulty lies in the mass or reduced mass that is found in many equations, including that for the rotational constant. We are dealing with molecular properties, not molar ones, and the mass required is that of a single molecule, given by the molar mass divided by N_A. The S.I. (Systéme Internationale) set of units requires that the molar mass be in kg mol^{-1}.

PROBLEMS

Any constants required are given at the end of the book.

2.1   Assume that $H^{35}Cl$ behaves as a rigid rotor and a Simple Harmonic Oscillator for which the rotational constant $B = 10.59\ cm^{-1}$ and $hv = 2885.98\ cm^{-1}$. By expanding the definitions of the partition functions (Equations 2.50 and 2.39) assess the number of levels that contribute significantly to them at 298 K, and evaluate the partition functions.

The ground electronic state is not split and the molecular orbitals lie at energies $\gg kT$. What is the electronic contribution to the total molecular partition function at this temperature?

2.2   Using B from Problem 2.1, calculate the approximate value of q_{rot} for $H^{35}Cl$ using Equation (2.53) and compare it with the accurate value calculated above. Comment on how good or bad the approximation is. Would it be as good for much lighter molecules?

2.3   Calculate the rotational constant of $^{35}Cl^{35}Cl$ in cm^{-1} given that the bond length is 198×10^{-12} m. Hence calculate the rotational contribution to the molecular partition function at 298 K.

(Use S.I. units throughout to obtain an answer in m^{-1}, and then convert it into cm^{-1}).

2.4   In reality molecules are not rigid rotors but have quantised energies given by

$$E_{rot} = BJ(J + 1) - DJ^2(J + 1)^2$$

where $D = 5.28 \times 10^{-4}\ cm^{-1}$ for HCl. For rotational levels $J = 0-10$ write down the corrected energy for each state and re-calculate their contribution to q_{rot} by direct summation. Compare the results with the rigid rotor ones in Problem 2.1. Is the difference significant in affecting the derived thermodynamic properties, such as the internal energy?

2.5   The rotational constants for $H^{35}Cl$, $F^{35}Cl$ and $I^{35}Cl$ are 10.59, 0.52 and 0.11 cm^{-1} respectively. What are the rotational temperatures, θ_{rot}, for each and what is their significance?

Using the vibrational frequency of $H^{35}Cl$ from Problem 2.1 and knowing that that for $F^{35}Cl$ is 783.6 cm^{-1}, calculate their characteristic vibrational temperatures, θ_{vib}, and comment on their implications.

2.6   Calculate the translational partition function of $H^{35}Cl$ molecules confined in a vessel of 100 cm^3 at 298 K and at 0.01 K.

2.7   Using information from the above problems, what is the value of the total partition function of $H^{35}Cl$ at 298 K?

CHAPTER 3

Thermodynamics

3.1 INTRODUCTION

Thermodynamics is a very sophisticated subject that has been carefully developed over time to describe and predict the physical properties of matter. It makes little sense to re-derive its well-established equations here and we shall use them without doing so. But in the foregoing chapters we saw the intimate relationship between the Internal Energy of a sample and the properties of the molecules that comprise it. Here we shall extend this and demonstrate how all thermodynamic functions are related to molecular properties through the partition function. In doing so we shall obtain additional insight into thermodynamics to that provided by the classical approach.

3.2 ENTROPY

In Section 2.1 we demonstrated that the derivation of the Boltzmann equation, which experiment shows accurately predicts the populations of energy levels under conditions of constant volume and thermal equilibrium, was based upon the existence of a dominant configuration of molecules between their energy states. In the canonical ensemble the value of the statistical weight W^* of the corresponding dominant configuration is conditional on the total number and total energy of the systems in the ensemble being constant. But there is no reason why the dominant configuration of one ensemble should be the same as that of another and it should not surprise us to discover that it is a characteristic of the ensemble. It tells us *how* the energy is distributed, and instinct makes us expect that an ensemble with one particular value of W^* might have different physical properties from another with a different value of it. But W^* is so large that it is unwieldy to use directly in this way and Boltzmann suggested that we should use its logarithm instead. He conse-

59

quently defined a new property, the *statistical entropy*, as

$$S = k\ln W^* \tag{3.1}$$

with k a proportionality constant. This is where the Boltzmann constant first entered science. It is apparent from its origins that S is related to the canonical partition function of the ensemble since both depend upon W^*.

The expression gives the total entropy of the whole ensemble of systems considered in the canonical ensemble, and to obtain it for one system we must divide by their number, N, as was done in the energy expression, Equation (2.23).

Taking all numbers n_i^* to be the populations of the systems in the dominant configuration of the ensemble,

$$S = \frac{k\ln W^*}{N} = \frac{k}{N}\left(\ln N! - \sum_i \ln n_i^*!\right) \tag{3.2}$$

which, using Stirling's approximation for the logarithms of factorials, becomes

$$S = \frac{k}{N}\left(N\ln N - \sum_i n_i^*\ln n_i^*\right) = \frac{k}{N}\sum_i n_i^* (\ln N - \ln n_i^*) = \frac{k}{N}\sum_i n_i^* \ln \frac{n_i^*}{N} \tag{3.3}$$

but,

$$\sum_i n_i^* = N \tag{3.4}$$

the total number of systems in the ensemble, and so,

$$S = -k\sum_i \ln \frac{n_i^*}{N} \tag{3.5}$$

From the canonical distribution,

$$\frac{n_i^*}{N} = \frac{e^{-\varepsilon_i/kT}}{Q} \tag{3.6}$$

so that

$$S = -k\ln\sum\left(\frac{e^{-\varepsilon_i/kT}}{Q}\right) = -k\left(\sum_i \frac{-\varepsilon_i}{kT} - \ln Q\right) \tag{3.7}$$

The sum of all the energies measured relative to the ground state defined as the energy zero is simply the total energy accepted by the system, which we have seen to be [U(T) − U(0)], giving the final result

$$S = \frac{U(T) - U(0)}{T} + k\ln Q \qquad (3.8)$$

The derivation may seem complicated, but the result is of fundamental importance. Q is related to the molecular partition function q and the properties of the molecules using the equations given in Chapter 2 for distinguishable or indistinguishable particles, as appropriate.

Throughout the derivation of $U(T)$ in Chapter 2 it was stressed that the volume of the ensemble was kept constant; only if this was true were our expressions for Q correct. It follows that in this expression for the entropy it is the value of $U(T)$ under conditions of constant volume that must be used (a different function is needed, under conditions of constant pressure, see below). We now must ask ourselves what is implied by a change in $U(T)$ due to a change in temperature at constant volume. Temperature does not change the energies of the quantised levels in the systems in the ensemble but rather their populations. This corresponds directly to a change in $U(T)$ $(=\Sigma_i n_i \varepsilon_i)$ but there is no change in Q $(=\Sigma_i g_i e^{-\varepsilon_i/kT})$, which is independent of level populations. It follows that for an infinitesimal increase dU in $U(T)$ the entropy increases as

$$S + \mathrm{d}S = \frac{U(T) + \mathrm{d}U - U(0)}{T} + k\ln Q = S + \frac{\mathrm{d}U}{T} \qquad (3.9)$$

Hence

$$\mathrm{d}S = \frac{\mathrm{d}U}{T} \qquad (3.10)$$

To accord with standard thermodynamic use we have dropped the explicit dependence of U on T, as we shall from here on. Under constant volume conditions this is precisely the classical thermodynamic definition of entropy, although it is usually stated in slightly different form (see below).

We now have equations that relate U and S to the canonical partition function. All other thermodynamic functions are defined in terms of these two quantities and so may themselves be expressed in terms of Q, but before we move on to this we consider some further basic principles.

3.3 ENTROPY AT 0 K AND THE THIRD LAW OF THERMODYNAMICS

In laboratory experiments we cannot always measure the absolute values of thermodynamic properties but only the change in them as some physical variable, such as temperature, is changed. To convert a measured change in the entropy of a system into an absolute value after the change has occurred requires us to know the absolute value at the initial temperature. Knowing the entropy at 0 K enables the absolute value at any other temperature to be obtained from a measurement of the change in entropy in attaining it. This is obtained from the 'Third Law' of thermodynamics, which we unusually introduce before the First and Second Laws.

The Third Law states that the entropy, $S(0)$, of a *perfect crystalline* solid at 0 K is zero and it follows from the statistical definition of entropy. By a perfect crystalline solid we mean, for example, a solid composed of atoms in which every identical atom sits exactly on its lattice site. The canonical ensemble then consists of n_0^* systems all in the same zero-eth energy state, and this number is therefore equal to the total number of systems, N. From the definition of statistical weight (see above) it follows that $W^* = 1$ and $\ln 1 = 0$, so that $S(0) = k \ln 1 = 0$. We therefore have our datum point for converting measured entropy changes into absolute values at any temperature. Note that the law is only true for a perfect crystalline solid, and not all solids are perfectly crystalline even at 0 K.

We note, too, that the derivation is wholly independent of what the crystal structure is and there is no difference in the entropies of perfect body-centred and face-centred cubic crystals, for example, at 0 K. Confusion often arises when allotropes are considered. It is, for example, experimentally possible to obtain values of the entropies of both monoclinic and rhombic sulfur at 0 K and it is tempting to think that one might have a different entropy from the other since it is the thermodynamically stable form at low temperature. This is entirely false reasoning since entropy is independent of energetic stability. All that matters is whether in each allotropic form the atoms are on the lattice points they should be *in its particular crystal structure*. Both allotropes have zero entropy at 0 K if in perfect crystalline form.

Whereas atoms are spherical, molecules may have shapes of lower symmetry and then a further requirement enters the definition of a perfect crystal. In a heteronuclear diatomic molecule such as CO, for example, the ends of the molecule are distinguishable from one another. Now not only must the molecule sit perfectly on its lattice point at 0 K but also all the molecules must be the same way round and point in the same direction. The perfect crystal consists of perfectly aligned CO molecules, with no OC ones intruding. Such perfect crystals are not found in practice (Section 3.5.2).

3.4 STATE FUNCTIONS

The two thermodynamic functions we have so far encountered, U and S, depend on the partition function and the absolute temperature, where the former is determined by the characteristic quantised energy levels in a given molecular system. It follows that for a specific molecular system at constant volume, where the energy levels are determined by molecular properties, their values change with temperature only. That is, at a given temperature, U and S for ensembles of defined, specific, molecules have values that are invariant. They depend solely on what the temperature is and are wholly independent of how the system was taken to that temperature. In turn this implies that if the temperature is changed, the changes in U and S are determined only by the start and finish temperatures and are completely independent of how the final temperature is attained from the initial one. We could, for example, accomplish the change in a single step, or we could proceed through any number of steps, increasing and decreasing the temperature at will to reach the final temperature. But the result, the measured change in U or S between the initial and final states, would be the same. We call U and S 'State Functions', meaning that their values depend simply on the physical state (here the volume and temperature) of the system.

The invariance of a State Function with path is a significant property that allows changes in them, resulting for example from a change in temperature, to be calculated very simply. It implies that if we go from one temperature to another infinitely slowly the magnitude of the change would be the same as if we changed the temperature instantaneously. We can therefore choose whatever (if necessary imagined) way of making the change we like to calculate it most easily, but we do not have to perform the change in the same way to measure it in the laboratory. We return to this point in Section 3.5.1. This property is so important that in classical thermodynamics all functions are defined to be State Functions. Here we have gained the insight that this is possible because they can all be expressed in terms of the energy levels and partition functions of the molecular assemblies they apply to.

3.5 THERMODYNAMICS

Thermodynamics was developed as a subject long before statistical thermodynamics was introduced, and before even the existence of atoms and molecules was generally accepted. Consequently, it does not consider them but instead provides a way for calculating the properties of large assemblies of them, and in particular changes in these properties as external conditions such as temperature and pressure are varied. But we have seen that these concepts are underpinned by quantum theory and statistics and we have discovered a powerful armoury for understanding the thermodynamic quantities in terms of the properties of the atoms and molecules involved. For many applications such insight is unnecessary, but we now know that if we wish to interpret the results of thermodynamic calculations at the molecular level we can do so. As a corollary we can appreciate how the precise natures of molecules affect their thermodynamic properties and, if we wish, synthesise them to have specific physical properties. In deriving our results we have seen that they are valid only for samples containing large numbers of molecules, and the same is implied for thermodynamics, too. Neither necessarily apply to experiments performed on small numbers of molecules nor to the devices and nanotechnology that exploit them.

This account is not intended as an adequate introduction either to thermodynamics or to statistical thermodynamics and the reader is advised to read it in conjunction with standard textbooks on these subjects. A deliberate decision has been made to focus on their inter-connections so as to unify the subjects. Here we shall emphasise the additional understanding that comes through the statistical, molecular, link.

3.5.1   First Law

The First Law is the law of conservation of energy. It states that the Internal Energy of a system is constant unless work (w) is done on it, or by it, or heat (q) is exchanged with the surroundings. Notably, we have explicitly allowed for both heat and work. Previously we have used U under conditions of constant volume so that no work, for example in expanding or compressing a gas, was involved, and our change in U above corresponds to a change in q here (the use of the same symbol for heat and the molecular partition function is an historic accident; care should be taken not to confuse them). For an infinitesimal change from this constant value,

$$dU = dq + dw \qquad (3.11)$$

This definition is quite general and applies to all possible sorts of work that a gas might do, or have done on it.

We consider a gas of volume V contained in a vessel subject to an external pressure, P, from a piston and at equilibrium with its surroundings. For this

$$dw = -PdV \qquad (3.12)$$

The negative sign has been introduced because, whereas the piston attempts to reduce the volume, the gas has to do equal and opposite work on it for equilibrium to be maintained. The gas loses energy by having to do this and U is decreased. This leads to the general convention that work done by a system is negative, and extends to any heat loss by the system being negative, too.

The concept of equilibrium is important and has a particular significance in thermodynamics and we expand what was said about changes above. If the pressure was to be infinitesimally increased then the volume would infinitesimally decrease, whereas if the pressure was similarly decreased the volume would also increase infinitesimally. A process in which the system responds in this way is termed 'reversible', and we say that an infinitesimal decrease in pressure causes a reversible expansion. This apparently obscure point allows, for example, the work done in a finite reversible expansion to be calculated, using calculus.

For an ideal gas $P = RT/V$ mol^{-1} and on expansion of the gas from a volume V_1 to V_2 the work done is

$$w = -PdV = -RT\int_{V_1}^{V_2}\frac{dV}{V} = -RT\ln\frac{V_2}{V_1} \tag{3.13}$$

That is, if we know V_1 we can calculate the work done in attaining the final volume. We note that the result depends on the initial and final values of the volume only, and we have discovered that volume is another State Function. Pressure is obviously also one.

We have seen above that the change in a State Function is independent of the way in which it occurs. Now we have discovered that if the change is performed in 'reversible equilibrium' way we can calculate it. However, to perform the change in this way would take infinite time since we would have to proceed in an infinity of infinitesimal steps at each one of which equilibrium was maintained. This is obviously impractical and in practice all measurements are made under *irreversible* conditions. In this example we might, for instance, suddenly jump the volume to its final value. But, since we are dealing with a State Function, the work done is the same as if we had proceeded infinitely slowly. This all seems very unreasonable when first met since it seems we apply calculations done under one set of conditions (reversible) to a system in which we accomplish the change under another set (irreversible), but this is precisely what makes the concept of State Functions so important. Here we glimpse the insight of the developers of thermodynamics in the 19th Century when they created such a subtle and powerful subject by defining their functions so carefully.

If a gas neither expands nor contracts in volume during a change then

$dw = 0$ and it follows from the First Law that $dU = dq_{rev}$, where we have written the subscript explicitly to remind ourselves of the reversible conditions that apply. We saw above that at constant volume $dS = dU/T$ and it follows that

$$dS = \frac{dq_{rev}}{T} \tag{3.14}$$

Although we have derived it under constant volume conditions the result is, in fact, quite general and applies to entropy changes under any conditions if the appropriate value of dq_{rev} is introduced. This generalised form is the classical thermodynamic definition of the entropy change in a process.

Most experimental work is performed at atmospheric pressure, and therefore under conditions of constant pressure rather than constant volume. We define another State Function, the enthalpy H as

$$H = U + PV \tag{3.15}$$

and for an infinitesimal change,

$$\begin{aligned} dH &= dU + PdV + VdP \\ &= dq_{rev} - PdV + PdV + VdP \\ &= dq_{rev} + VdP \end{aligned} \tag{3.16}$$

where we have assumed equilibrium reversible conditions and the gas has done work $dw = -PdV$ against the external pressure. For a gas at constant pressure the final term is zero and $dH = dq_{rev}$ so that

$$dS = \frac{dH}{T} \tag{3.17}$$

As with U there may be no way of measuring H directly, and we once more use the heat capacity, now measured at constant pressure, C_P. It is defined as

$$C_P = \left(\frac{\partial H}{\partial T}\right)_P \tag{3.18}$$

and it follows that

$$dS = \frac{C_P dT}{T} \tag{3.19}$$

This is the expression for an infinitesimal entropy change within a pure phase, a solid for example, caused by an infinitesimal change in the temperature. But if a phase change occurs to change the solid to a liquid then the enthalpy change associated with it at the transition temperature T_{trans} is the enthalpy of transition, ΔH_{trans} (which can be measured directly) and the entropy change of the transition at this temperature is

$$dS_{\text{trans}} = \frac{\Delta H_{\text{trans}}}{T_{\text{trans}}} \tag{3.20}$$

These are important relations that allow us to calculate entropy changes in real situations and thereby to obtain the absolute entropy since the absolute entropy at 0 K is known through the Third Law.

3.5.1.1 Absolute Entropies and the Entropies of Molecular Crystals at 0 K. We consider the simplest situation in which a material exists with only one solid phase. We heat a perfect crystal until it melts and becomes liquid at the fusion temperature, T_{fus}, after which further heating causes it to boil and become gaseous at the boiling point, T_{boil}. The gas is then heated to its final temperature T:

perfect crystal at 0 K $\xrightarrow{T_{fus}}$ liquid $\xrightarrow{T_{boil}}$ gas → gas at temperature T (3.21)

Thermodynamics then tells us that the absolute value of the entropy at T is:

$$S = 0 + \int_{0}^{T_{fus}} \frac{C_{\text{p}}(\text{solid})dT}{T} + \frac{\Delta H_{\text{fus}}}{T_{\text{fus}}} + \int_{T_{fus}}^{T_{boil}} \frac{C_{\text{p}}(\text{liquid})dT}{T} + \frac{\Delta H_{\text{vap}}}{T_{\text{boil}}} + \int_{T_{boil}}^{T} \frac{C_{\text{p}}(\text{gas})dT}{T}$$

(3.22)

where we have simply added each contribution in turn. Its evaluation depends on measuring the (different) heat capacities of each of the three phases over the temperature range they exist, and also the enthalpies of fusion (ΔH_{fus}) and vapourisation (ΔH_{vap}). In detail the heat capacity of the solid must be extrapolated to 0 K using the Debye Law ($C_{\text{p}} \propto T^3$) to allow the first integration. The final result is known as the 'thermodynamic entropy' of the gas at T.

However, we have seen in Chapter 2 that there is a completely independent way of assessing the entropy using statistical thermodynamics and the canonical partition function. This requires knowledge of the energy levels of the molecules concerned and the result is known as the 'spectroscopic' or 'statistical' entropy. If all our assumptions are correct then the two calculations should give the same value. This provides an invaluable check on the Third Law since only if it is correct will this be true. For a perfect crystal consisting of atoms, such as solid argon, it is. If the crystal contained dislocations and was not perfect it would not be.

A corollary is that comparison of the two values allows us to assess whether a crystal is perfectly ordered at 0 K and we return, in particular, to a molecule with a non-spherical shape with distinguishable ends, for example the heteronuclear diatomic molecule CO. Measurements on samples of CO reveal a discrepancy between the two values, which can be explained only by concluding the solid to have a non-zero entropy of 5 J K^{-1} mol^{-1} at 0 K; such a non-zero value is known as a 'residual entropy'. The explanation lies in two different factors contributing to the overall order in the crystal. The first is whether all the molecules sit on their exact lattice points and the second whether they are all similarly arranged (*i.e.* point in the same direction) on these points. The value of the most probable statistical weight W^* is then determined by the product of separate values describing the most probable statistical weights of the two. In the perfect CO lattice at 0 K all the molecules are indeed on their lattice points and the factor contributing to W^* remains 1, but some of the molecules point in one direction and the others in the opposite one. The crystal is therefore not perfect in the sense described in Section 3.3 and residual disorder remains at 0 K.

If the two orientations are equally likely, in 1 mole there are 2^{N_A} ways of arranging the molecules on their lattice sites, making the total $W^* = 1 \times 2^{N_A}$ so that the entropy would be expected to be

$$S = k\ln(1 \times 2^{N_A}) = R\ln 2 = 5.76 \text{ J K}^{-1} \text{ mol}^{-1} \qquad (3.23)$$

This is nearly equal to the observed value but the discrepancy is real and implies that the two orientations are not actually equally likely – in the crystal the two ends of the molecule are not quite randomly ordered.

Remarkably, only two orientations exist since the crystal grows in the laboratory by molecules randomly hitting its cold surface and we might expect that at very low temperatures many orientations might be possible since the molecules might freeze in any arbitrary direction they land in. Apparently intermolecular forces make the surface essentially grooved in its energy profile so that the molecules lie either along the groove or against it. The disorder is frozen in at very low temperatures because the molecules on the surface have insufficient energy to overcome the potential barrier between the two orientations.

With a homonuclear diatomic molecule such as $^{16}O-^{16}O$ we cannot distinguish between the ends of the molecule and no such disorder can exist, so that the residual entropy is zero in a perfect crystal. The same is true for any symmetric linear molecule and it is an interesting historic fact that N_2O was thought to have a symmetric structure until the existence of a residual entropy revealed it to have the structure NNO.

The arguments we have used here apply to all molecules, and they may have different residual entropies as a result. For example, a mono-substituted tetrahedral methane CH_3X has four distinguishable orientations on a lattice point, and a residual entropy near $R\ln4$ J K^{-1} mol^{-1}; a monosubstituted hexagonal benzene molecule would have six orientations and a residual entropy near $R\ln6$ J K^{-1} mol^{-1}.

Finally, whilst only one gas phase exists in any chemical system, and only one liquid one (except for He), many systems can exist in a number of solid phases of different crystal structure or ordering, and therefore exhibit many solid-state phase changes as the temperature is changed. The treatment we have given for calculating the thermodynamic entropy is simply extended to these cases by adding appropriate terms for each, and for the entropy changes in the different solid phases between the transition temperatures (Problem 3.4).

3.5.2 Second Law

The Second Law is concerned with the entropy of systems and is a law of common sense. But, as ever in thermodynamics, we must define our terms carefully to ensure that we understand what we mean by this. An important concept is that of *spontaneous change* and is best illustrated by an example. If we confine an atomic gas in a vessel but then connect it to another of similar volume in which there is initially a vacuum we know that the atoms spontaneously flow out of the first and disperse themselves over the whole new volume. In a system maintained at constant temperature the total energy of the gas is unchanged since this depends only on the temperature, and the Maxwell distribution is completely independent of the volume, so the process is not driven by energetic considerations. But what about the converse? Do we expect all the atoms in the new large volume suddenly all to go back where they came from? Although this is theoretically possible experience shows that it never happens and return to the starting condition is not spontaneous. The expansion is irreversible.

The spontaneous change in the closed system (that is one no atoms either leave or enter) occurs so as to increase the randomness of the gas. Before the expansion we knew that all the atoms were in the initial vessel, but after it they are distributed over both and we have less knowledge of where exactly they are. When discussing the canonical ensemble we were careful to conserve its total volume but here the volume has doubled. The dominant configuration in the doubled volume is different from that in the initial volume, and corresponds to an increase in the statistical weight

W^* and, consequently, to an increase in the entropy. This is general and leads to a simple rule. Any change in randomness implies a change in entropy and we say that spontaneous changes are 'entropy driven'. All irreversible processes, such as gas expansion, are spontaneous and lead to an increase in entropy in an isolated system. But reversible changes do not.

The Second Law is a statement of this: 'In an isolated system the entropy increases in a spontaneous change'.

This is all we need for our discussion here, although it should be emphasised that the full implications of the Second Law, covered in standard thermodynamic textbooks, are many and various.

Now we have some insight into what, from a thermodynamic view point, determines whether any chemical or physical process might occur. We often have to supply energy to a system to effect chemical or physical change. But in entropy we have discovered another factor that must be considered and the Second Law must be obeyed at the same time. We have therefore to consider the energy and entropy requirements for processes to occur in tandem. This is accomplished through the concept of 'Free Energy'.

3.6   FREE ENERGY

We define two new State Functions, the Helmholtz (A) and Gibbs (G) Free Energies to include both energy and entropy. In a system at constant volume,

$$A = U - TS \tag{3.24}$$

whereas in one at constant pressure,

$$G = H - TS \tag{3.25}$$

In both cases the factor T is introduced to ensure dimensional correctness (*i.e.* all terms are energies). At constant temperature the finite changes in Free Energy in a process are consequently

$$\Delta A = \Delta U - T\Delta S \tag{3.26}$$

and

$$\Delta G = \Delta H - T\Delta S \tag{3.27}$$

respectively. In each the term containing the entropy adds to or subtracts from an energy term. For a spontaneous process ΔS is positive in an isolated system and, if ΔU or ΔH is positive, the entropy term decreases the overall energy change that occurs. What remains is the 'free' energy left to accomplish change. This is a simplistic argument since in practice any of the energy or entropy quantities may be positive or negative, according to the precise system under study.

One can think of the entropy term as providing a source of difficulty for extracting the energy since it represents the dispersion of energy through-out the system. This means that the available energy will always be less than the maximum represented by the internal energy or the enthalpy. An analogy might be robbing a bank or banks of 1 million pounds. If all the money is in one bank one only has to rob this one, which may be difficult but is feasible. But if the money is distributed equally through 1 million banks it becomes effectively impossible to get one's hands on it all.

The properties of Free Energy, and its implications to change or equilibrium in isolated systems have long been established and are of supreme importance in thermodynamics. Some evidence for their correct-ness is given in a discussion of the freezing of water, below. They may be summarised:

(1) The change in Free Energy during a process, ΔA at constant volume or ΔG at constant pressure, is negative in a spontaneous process.
(2) The condition for equilibrium between two phases is that their Free Energies are equal. This implies that the change ΔA or $\Delta G = 0$.
(3) If a Free Energy change in a process is positive then it is ther-modynamically forbidden and cannot happen.

These rules apply to pure substances whereas chemistry usually involves mixtures of different types of molecules. To cope with this a new quantity is defined. This is the *partial molar Free Energy*, also known as the *chemical potential* and, when working under constant pressure condi-tions, is given for each component, i, by

$$\mu_i = \left(\frac{\partial G}{\partial n_i}\right)_{n_j, T, P} \tag{3.28}$$

That is, it is the Gibbs Free Energy change at constant T and P when a change in the concentration of component i occurs without any change in the concentration of any of the other species, j, present. Now $\Delta \mu_i$ must be negative for a spontaneous change, and $\Delta \mu_i = 0$ at equilibrium, whilst if $\Delta \mu_i$ is positive the change cannot occur. Under conditions of constant volume a different chemical potential is defined:

$$\mu_i = \left(\frac{\partial A}{\partial n_i}\right)_{n_j, T, V} \tag{3.29}$$

The thermodynamic rules tell us if a change *could* occur, not that it *will* occur. For example, for water at 273 K and 1 bar pressure the Gibbs free energy change to form ice, ΔG, is zero, and ice and water are in equilibrium. For this phase change ΔH is negative, since heat has to be lost for freezing to occur. But the entropy change at the phase change is given by $\Delta H / T_{fusion}$ and since in the definition of ΔG this is multiplied by $-T_{fusion}$ the term is positive and ΔG is identically zero, as stated. From experience at lower temperatures freezing may occur and we note that $T\Delta S$ is then less than ΔH, making ΔG negative. Change then becomes allowed as a spontaneous process in the thermodynamic sense. But this phase change may never happen because change is a kinetic process, and an activation energy barrier must be overcome for it to occur. The word 'spontaneous' has very different meanings in thermodynamics and in normal life. Pure water can in fact be maintained at temperatures lower than 273 K indefinitely without freezing. That water normally does freeze at 273 K is due to the crystallisation being catalysed by specks of dust that provide nucleation centres for ice and a catalysed low activation energy route for the process to occur. However, if we calculate the free energy change at temperatures above 273 K we discover ΔG to be positive, and experience tells us that water does not freeze above this temperature. We conclude that a positive value indicates that a process is impossible. These conclusions are all consistent with the rules governing free energy changes given above. Furthermore, the example establishes the important principle that thermodynamics can tell us what processes are possible under given conditions but is completely silent on whether they do happen, and certainly does not imply that they *must* happen. If a change does occur, though, thermodynamics decrees the direction of it.

We now return to the expression of all thermodynamic quantities in terms of partition functions to allow us insight at a molecular level into what controls change. Our world increasingly exploits the quantum nature of matter in computers, superconductors, photovoltaic cells, *etc.* and our approach is necessary to understand these and to design new devices for specific purposes.

3.7 THERMODYNAMIC FUNCTIONS AND PARTITION FUNCTIONS

We start from the two equations we have already derived and use thermodynamic definitions to derive the rest. Since

$$S - S(0) = \frac{U - U(0)}{T} + k\ln Q \qquad (3.30)$$

where $S(0) = 0$ for a perfect crystalline solid, it follows that

$$A - A(0) = [U - U(0)] - TS = -kT\ln Q \qquad (3.31)$$

This leads to a simple expression for the chemical potential defined under conditions of constant volume and constant temperature. From above,

$$\mu = \left(\frac{\partial A}{\partial n_i}\right)_{n_J,V,T} = N_A\left(\frac{\partial A}{\partial N_i}\right)_{n_J,V,T} = -RT\left(\frac{\partial \ln Q}{\partial N_i}\right)_{n_J,V,T} \qquad (3.32)$$

Combining the First and Second Laws, classical thermodynamics tells us that for an infinitesimal change of A in a closed system

$$dA = -PdV - SdT \qquad (3.33)$$

and, therefore,

$$P = -\left(\frac{\partial A}{\partial V}\right)_T = kT\left(\frac{\partial \ln Q}{\partial V}\right)_T \qquad (3.34)$$

Now we use the definition of the enthalpy to obtain it in terms of Q:

$$H - H(0) = [U - U(0)] + PV = kT^2\left(\frac{\partial \ln Q}{\partial T}\right)_V + kT\left(\frac{\partial \ln Q}{\partial V}\right)_T \qquad (3.35)$$

Similarly,

$$G - G(0) = [A - A(0)] + PV = -kT\ln Q + kTV\left(\frac{\partial \ln Q}{\partial V}\right)_T \qquad (3.36)$$

All of these equations are perfectly general and apply to interacting and non-interacting molecules. We must, as already seen, write $Q = q^N$ if the particles are distinguishable and $Q = q^N/N!$ if they are not; this may seem complicated but the choice just requires common sense when applied to real problems (Chapter 4). However, these are the crucial equations through which the fact that the energy levels of molecules are quantised enters thermodynamics.

For a perfect gas, since $PV = nRT$, we may write G in an alternative form,

$$G - G(0) = [A - A(0)] + PV = -kT\ln Q + nRT \qquad (3.37)$$

and for a gas $Q = q^N/N!$ so that

$$G - G(0) = -NkT\ln Q + kT\ln N! + nRT$$
$$= -nRT\ln q + kT(N\ln N - N) + nRT = -nRT\ln(q/N) \tag{3.38}$$

Here we have once more used Stirling's approximation for the logarithm of a factorial. Within chemical problems it is often convenient to express the Gibbs Free Energy in terms of a *molar partition function*, q_m, defined as q/n mol^{-1}:

$$G - G(0) = -nRT\ln(q_m/N_A) \tag{3.39}$$

3.8 CONCLUSION

We now have the armoury to calculate all the thermodynamic properties of ensembles of molecules from a knowledge of the quantised energy levels of the molecules that comprise them and which depend on the nature and structures of these molecules. We have simply to insert the expressions containing partition functions derived above into the long-established thermodynamic equations. In doing this we have gained continuity in that all of our instincts and training are to think of chemistry in terms of what happens at molecular level, and we have brought the seemingly esoteric field of thermodynamics into the same compass. We shall also see in the next chapter that our approach gives us real insight into what governs the behaviour we calculate thermodynamically and see experimentally, and through it we can design our experiments and technology better. Using our expressions is very straightforward, although at first sight the calculations may seem long-winded. But all are based on a few basic equations that we use time and time again.

PROBLEMS

Any constants required are listed at the end of the book.

3.1 The basis of statistical thermodynamics lies in the distributions of molecules between their quantised energy states and we have invoked some important standard results in the text. Here we check that these are reasonable by showing in a very small system that some distributions are more likely than others. Take 4 particles to be distributed over 4 equally probable states that differ in energy. By labelling them A, B, C and D so that they can be distinguished verify that a distribution containing 2 particles in the first level, 0 in the next, and 1 in each of the others can be attained in 12 ways, whilst the distribution with, for example, 3 in the first and 1 in the third, with the others empty, can be obtained in only 4 ways.

Further, verify that their statistical weights are given by $W_1 = 4!/(2!0!1!1!)$ and $W_2 = 4!/(3!0!1!0!)$.

These are obviously just two of many different possible distributions. If this calculation is performed using much greater numbers of particles and levels it is found that the statistical weight W of one distribution (the most probable) becomes very much greater than of any other.

3.2 The heat capacity at constant pressure of a gas is found empirically to vary with temperature according to the expansion

$$C_p(T) = A + BT + CT^2$$

where A, B and C are temperature-independent constants. The change in entropy in heating the gas from temperature T_1 to T_2 is given by

$$\Delta S = \int_{T_1}^{T_2} \frac{C_p(T)\mathrm{d}T}{T}$$

Obtain a general expression for this quantity.

For Cl_2, and with $C_p(T)$ in J mol^{-1} K^{-1}, $A = 31.70$, $B = 0.001$ and $C = -2.7 \times 10^{-7}$. Calculate the entropy change in heating Cl_2 from 300 to 900 K.

3.3 From Table 3.1 calculate the entropy change of fusion and vapourisation for the substances given.

Table 3.1

	ΔH_{fus} (kJ mol^{-1})	T_{fus} (K)	ΔH_{vap} (kJ mol^{-1})	$T_{boiling}$ (K)
N_2	0.72	63.3	5.58	77.4
NH_3	5.65	195.4	23.3	239.7
H_2O	6.01	273.2	40.7	373.2

In each case the entropy of vapourisation is greater than that of fusion. Why is this what you would expect?

3.4 At 0 K dinitrogen forms a solid of a particular structure but as it is warmed it undergoes a solid-state phase transition at 35.6 K and adopts a different crystal structure. Further warming causes it first to become a liquid and then a gas. Given the following information, calculate the absolute entropy of N_2 at 298 K.

$T_{transition} = 35.6$ K, $T_{melting} = 63.1$ K, $T_{vapourisation} = 72.1$ K

$\Delta H_{transition} = 228.9$ J mol^{-1}, $\Delta H_{melting} = 721.1$ J mol^{-1}, $\Delta H_{vapourisation} = 5200.5$ J mol^{-1},

$$\int_0^{35.6} \frac{C_p(\text{solid } 1)dT}{T} = 27.2 \text{ J K}^{-1} \text{ mol}^{-1}; \int_{35.6}^{63.1} \frac{C_p(\text{solid } 2)dT}{T} = 23.4 \text{ J K}^{-1} \text{ mol}^{-1}$$

$$\int_{63.1}^{72.1} \frac{C_p(\text{liquid})dT}{T} = 11.41 \text{ J K}^{-1} \text{ mol}^{-1} \int_{72.1}^{298} \frac{C_p(\text{gas})dT}{T} = 39.2 \text{ J K}^{-1} \text{ mol}^{-1}$$

3.5   Sulfur can exist in two allotropic forms. Monoclinic sulfur is the stable form at higher temperatures but it may undergo a transition to the rhombic form at 369 K. That is, it becomes thermodynamically possible for this change to occur at this temperature. Since change is a kinetic process, however, the transition may not occur and in fact the heat capacities of both forms can be measured down to near 0 K, with the values at the lowest temperatures estimated using the Debye extrapolation.

From integrating the heat capacity data we find that $S_{\text{mono}}(369$ K$) = S_{\text{mono}}(0 \text{ K}) + 37 \text{ J K}^{-1} \text{ mol}^{-1}$ and $S_{\text{rhomb}}(369 \text{ K}) = S_{\text{rhomb}}(0) + 36$ J K^{-1} mol^{-1}. The enthalpy change in the transition is 402 J mol^{-1}.

Show that these figures are consistent with the Third Law of thermodynamics.

(Hint: remember that entropy is a state function so that the entropy of monoclinic sulfur at 369 K is independent of the route taken to reach this point).

3.6   Calculate the statistical entropy of 1 mole of $^{16}O_2$ contained in a 100 cm^3 vessel at 298 K given that the rotational constant $B = 14.37$ cm^{-1} and neglecting any contribution to the partition function from vibrational levels above the first, which we take as the zero of energy. (Higher electronic levels make a completely negligible contribution.) Use the Equipartition value for $U(T)$ at this temperature.

What would you expect the residual entropy (that at 0 K) to be for ^{16}O–^{16}O and for ^{16}O–^{18}O?

CHAPTER 4

Applications

4.1 GENERAL STRATEGY

All thermodynamic functions can be expressed in terms of the canonical partition function, and through it the molecular partition function. This may always be calculated from measurement of the quantised energy levels in systems, without approximation. However, real chemical insight into what controls the chemical and physical properties of materials can be gained by using the usually very good approximation that the molecular partition function can be written as the product of individual factors representing the partition functions for translational, rotational, vibrational and electronic energy. This is because the use of specific models of the motions involved allows us to express the energies in terms of the structures of the molecules themselves.

In this chapter we shall substitute the expressions for these derived in Chapter 2 into some classical thermodynamic equations. Since we may in general deal with four factors to the partition functions, originating in the different contributions to the energy, the expressions may appear long, but no new mathematics is involved over that already used. It is just a matter of being systematic and using the same expressions time and time again in different chemical and physical situations. We illustrate this with a series of examples.

4.2 ENTROPY OF GASES

Statistical thermodynamics allows our ideas of entropy to be sharpened. Entropy has long been associated with the order or disorder in systems, but in the absence of statistics order is not a very clearly defined property. For example, if we take a sample of an inert monatomic gas, helium, say, at 1 bar pressure in a given volume at room temperature we would recognise that the atoms are disordered as a result of their translational motion and their frequent collisions. With this almost ideal gas, any tendency to ordering through intermolecular forces is negligible. So if we replace the helium with neon, with similar properties, we might expect the atoms to be equally disordered and the entropies of the two samples to be the same. Experimentally this is not found to be so and we seek an explanation though our statistical thermodynamic approach.

4.2.1 The Entropies of Monatomic Gases

The significance in commencing our discussion of entropy with monatomic gases is that most normally possess only translational energy, which simplifies the problem. We start from our general expression for the entropy of 1 mole of a substance

$$S = \frac{U - U(0)}{T} + kT\ln Q_{trans} = \frac{U - U(0)}{T} + RT\ln q_{trans} - k\ln N_A! \quad (4.1)$$

since the molecules are indistinguishable, and here $N = N_A$ and $kN_A = R$. Using Stirling's approximation for $\ln N_A!$ and knowing that the translational energy of atoms at temperatures above very close to 0 K is equal to that predicted by Equipartition Theory (Chapter 1), and $U(0) = 0$, $U - U(0) = \frac{3}{2}RT \, mol^{-1}$ gives

$$S = \frac{3R}{2} + R(\ln q_{trans} - \ln N_A + 1) \quad (4.2)$$

This is normally re-written as

$$S = R(\ln e^{3/2} + \ln q_{trans} - \ln N_A + \ln e) = R\ln\left(\frac{e^{5/2} q_{trans}}{N_A}\right) \quad (4.3)$$

Now we substitute our expression for q_{trans} and introduce the molar volume, V_M:

$$S = R\ln\left[\frac{e^{5/2} V_M}{N_A}\left(\frac{2\pi mkT}{h^2}\right)^{3/2}\right] \quad (4.4)$$

This is the Sackur–Tetrode equation.

Clearly, the entropy depends on the atomic mass, something we had no idea of before we used the statistical thermodynamic approach. The calculated values for the standard molar free entropies of Ne and Ar at 298 K are 17.60R and 18.62R respectively (146.5 and 155.0 J K^{-1} mol^{-1}) and are in good agreement with experiment (Problem 4.1).

4.2.2 Entropy of Diatomic and Polyatomic Molecules

The calculation of translational entropy gives the correct value for its contribution to the entropy of molecules too, but these may also have rotational and vibrational contributions. However, this is very simply dealt with since, under our normal approximation of separability of energies,

$$Q = Q_{trans}Q_{rot}Q_{vib} \qquad (4.5)$$

and

$$U = U_{trans} + U_{rot} + U_{vib} \qquad (4.6)$$

It immediately follows that the entropy can also be expressed as the sum of different contributions.

$$S = S_{trans} + S_{rot} + S_{vib} \qquad (4.7)$$

We have seen that, near room temperature, diatomic molecules are in their ground vibrational states, and with reference to this energy state $U_{vib} = 0$ and $Q_{vib} = q_{vib}^N = 1$ since $q_{vib} = 1$, so that $S_{vib} = k\ln q_{vib} = 0$. We therefore need only to calculate the extra contribution due to rotation. This may, of course, be untrue at higher temperatures where further vibrational levels may be occupied.

We have seen also that at room temperature the rotational contribution to the energy of a diatomic molecule is given by the Equipartition value, $U_{rot} = RT$ J mol^{-1} and $U_{rot}(0) = 0$ so that

$$S_{rot} = R + kT\ln Q_{rot} = R + k\ln q_{rot}^{N_A} = R + R\ln\left(\frac{kT}{\sigma hcB}\right) \qquad (4.8)$$

Note that for both rotation and vibration the ends of the molecule are distinguishable so that $Q = q^N$, as we have stated previously.

The contribution to the entropy depends upon whether the molecule is homonuclear ($\sigma = 2$) or heteronuclear ($\sigma = 1$), and also upon the masses of the atoms through the value of the rotational constant, B (Equation 2.48). This is inversely proportional to the reduced mass so that diatomics composed of heavy atoms have greater contributions to the rotational entropy at a given temperature than those consisting of light ones. As with the translational contribution, atomic masses affect the rotational entropies of gases although they do not affect the internal energy contributions, once the Equipartition condition has been met, as the temperature is raised.

The overall entropy of diatomic molecules clearly exceeds that of atoms, and varies from one specific chemical type of molecule to another.

Extension to more complicated cases is straightforward, using the separability of contributions. For example, a non-linear molecule has three separate contributions to its rotational entropy through having three different moments of inertia, and three different contributions to the partition function. A vibrating polyatomic molecule such as CO_2 may have in addition four separate (two equal) contributions to its vibrational entropy at a sufficiently high temperature at which all the stretching and bending motions are excited. In general this occurs at lower temperatures than in diatomics because in heavier molecules the vibrational levels are closer in energy.

4.3   TWO-LEVEL SYSTEMS; ZEEMAN EFFECTS AND MAGNETIC RESONANCE

The significance of two-level systems to our understanding was outlined in Chapter 1, where it was pointed out that some important real examples exist. Here we simply put the discussion into a general context.

4.3.1   Internal Energy of Two-level Systems

At thermal equilibrium it is trivial to calculate the Internal Energy of the system. We define the lower energy level to have energy 0 and the higher one energy ε, (Figure 1.8). For such a simple case it is simplest not to use the partition function at all:

$$U = n_0 \times 0 + n_1 \times \varepsilon = n_1 \varepsilon \qquad (4.9)$$

Using the Boltzmann distribution for 1 mole,

$$\frac{n_1}{N_A} = \frac{e^{-\varepsilon/kT}}{e^0 + e^{-\varepsilon/kT}} = \frac{e^{-\varepsilon/kT}}{1 + e^{-\varepsilon/kT}} \qquad (4.10)$$

and hence

$$U = N_A \varepsilon \frac{1}{(e^{\varepsilon/kT} + 1)} \qquad (4.11)$$

Noting that the denominator in the Boltzmann equation is in fact the partition function we may obtain the same result using the equations we have developed:

$$U = N_A kT^2 \frac{d \ln q}{dT} = N_A kT^2 \frac{d \ln(1 + e^{-\varepsilon/kT})}{dT}$$

$$= N_A kT^2 \frac{\varepsilon}{kT^2} \left(\frac{e^{-\varepsilon/kT}}{1 + e^{-\varepsilon/kT}} \right) = N_A \varepsilon \left(\frac{1}{e^{\varepsilon/kT} + 1} \right) \qquad (4.12)$$

This is readily evaluated if ε is known. For example, for electrons in a magnetic field, B, the energy of the higher state above the lower is

$$\varepsilon = g\mu_B B \qquad (4.13)$$

where g is the 'g-factor' (2.0023 for the free electron) and μ_B is the primary unit of atomic magnetism, the Bohr magneton.

It is similarly straightforward to calculate the entropy of the system using the partition function and the value for U (although see below for a convenient choice of energy origin).

4.3.2 Curie Law

So far we have shown how to calculate energies and entropies but our methods allow us to calculate *any* physical property of the system. We extend what was done in Chapter 1 to discuss this in the case of the magnetism of a two-level system of molecules containing single unpaired electrons (free radicals) placed in an external magnetic field. Since the electrons in the two spin states have different energies due to being magnetic in opposite senses, and since the two energy levels are unequally populated, the sample has a bulk magnetism. The magnetic moments are given by $\mu = m_S g\mu_B$, where $m_S = \pm\frac{1}{2}$ for the electron.

For each state we calculate the magnetism associated with all the electrons that occupy it by multiplying these individual values by the populations of the states, and then add them to obtain the bulk magnetisation of the sample, M:

$$M = \tfrac{1}{2}g\mu_B n_0 - \tfrac{1}{2}g\mu_B n_1 \qquad (4.14)$$

For 1 mole of electrons, using the Boltzmann distribution we obtain

$$n_0 = N_A \left(\frac{e^{-1/2g\mu_B B/kT}}{e^{-1/2g\mu_B B/kT} + e^{1/2g\mu_B B/kT}} \right) \qquad (4.15)$$

and similarly for n_1. Note that here we have departed from our usual rule of putting the energy of the lowest level equal to zero and have chosen instead to make the lowest level in the *absence* of the field zero (Figure 4.1), something which experience shows eases solution of problems such

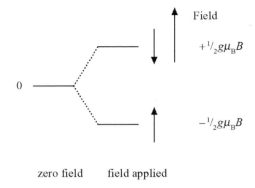

Figure 4.1 *Magnetism of a two-level system. The lowest level is chosen as that in the absence of the field.*

as this (it also eases calculating the entropy of such a two-level system). The energies with respect to this zero are $\pm\frac{1}{2}g\mu_B B$. Inserting the values of the populations gives

$$M = \frac{N_A g \mu_B}{2}\left(\frac{e^{1/2 g\mu_B B/kT} - e^{-1/2 g\mu_B B/kT}}{e^{1/2 g\mu_B B/kT} + e^{-1/2 g\mu_B B/kT}}\right) \tag{4.16}$$

For all normal field strengths the exponents are $\ll 1$ and we expand the exponentials, truncating them after the first term:

$$M \approx \frac{N_A g \mu_B}{2} \times \frac{g\mu_B B}{2kT} = \frac{N_A (g\mu_B)^2 B}{4kT} \equiv \chi B \tag{4.17}$$

where χ (Greek chi) is defined as the molar bulk susceptibility and is a measurable quantity. It can be expressed as

$$\chi = \frac{\text{const}}{T} \tag{4.18}$$

a relation found empirically by Pierre Curie and named after him. It is a valuable relation in inorganic chemistry and in materials science and our derivation has shown its underlying physical origin. The simple connection between the magnetisation and the applied field strength underlies many modern magnetic devices.

4.4 THE INTENSITIES OF SPECTRAL LINES

Spectroscopic transitions are excited between two quantised energy levels according to the Bohr relation, Equation (1.8). As mentioned previously, this does not decree whether a molecule absorbs radiation or emits it under the influence of the incident light and in fact transitions may occur in both directions when recording a spectrum. Einstein showed early on that the rate *constant* (for which he used the symbol B; unfortunately this symbol is also used for magnetic field strength, above) for a molecule in the lower state to absorb energy was exactly equal to the rate constant for one in an upper energy state to emit energy. It follows that the net rate of absorption of energy from the incident light beam, the intensity of the transition, is given by the difference in the *rates* (rate constant times population) of absorption and emission:

$$I = Bn_{\text{lower}} - Bn_{\text{upper}} \tag{4.19}$$

where the n's are the populations of the levels.

If the transition is allowed by spectroscopic selection rules B is non-zero and the intensity depends on the population *difference* between the two levels. Provided that the molecules are in thermal equilibrium with their surroundings, this gives the opportunity to verify the Boltzmann equation experimentally. We are careful to stipulate this since in many modern devices, such as lasers, this condition is deliberately broken.

We have seen that at room temperature it is only the lowest vibrational level that is populated, but many rotational ones are. In the infrared spectroscopy of diatomic molecules (Figure 2.6) transitions are excited between these occupied rotational levels and the unoccupied ones associated with the next lowest vibrational level, according to the selection rules $\Delta v = 1$, $\Delta J = \pm 1$. The intensities of the spectral lines therefore provide a measure of the relative populations of the rotational levels in the lower vibrational level, the upper one being empty. In saying this we have assumed that the transition probability, and the value of the rate constant, is independent of which precise rotational levels the transition connects; in fact a correction should be made before comparison with the Boltzmann predictions is made.

The assumption is most dangerous when transitions are observed (in the microwave region) between the rotational levels themselves. For these, quantum mechanical calculations show that the rate constants, B, are proportional to $(J + 1)/(2J + 1)$. This causes the rate of absorption of energy, for example, to be proportional to $(J + 1)$ rather than $(2J + 1)$:

$$\text{rate of absorption} \propto \frac{(J + 1)}{(2J + 1)} n_0 = \frac{(J + 1)}{(2J + 1)} (2J + 1) e^{-J(J + 1)\frac{\theta_{\text{rot}}}{T}}$$

$$= (J + 1) e^{-J(J + 1)\frac{\theta_{\text{rot}}}{T}} \tag{4.20}$$

but, once correction is made for this, agreement with experiment is excellent for a thermally equilibrated sample. Either the infrared or the microwave spectrum of a sample can be used to measure its temperature by fitting the observed spectral intensities to the theoretical equations.

The primary requirement for a molecule to display an infrared spectrum is that it must have a permanent electric dipole moment to interact with the electric vector of the light beam for pure rotational transitions, or a dipole fluctuating at the vibrational frequency for infrared ones. Neither is possessed by homonuclear diatomic molecules such as $^{16}O-^{16}O$, and they consequently have no IR spectra. But they do exhibit Raman spectra that again allow the populations of the levels to be monitored, although now pure rotational transitions correspond to changes of ± 2 in J. But from these spectra we discover the important fact that in this molecule, for example, every alternate rotational state is missing – a consequence of the Pauli Principle applied to diatomics composed of nuclei that do not possess spin angular momentum. This has an obvious direct influence on the thermodynamic properties of samples of such molecules too.

4.5 PAULI PRINCIPLE AND *ORTHO* AND *PARA* HYDROGEN

The Pauli principle has its origins in fundamental quantum mechanics and tells us that there are symmetry restrictions on the wave function of a *symmetric* molecule. The wave function is a complete description of the molecule and so if the molecule is composed of nuclei with spin angular momenta ('nuclear spin, ns') then this information must be contained within it. ^{1}H is a spin-$\frac{1}{2}$ nucleus whereas ^{16}O has no spin. By quantum laws two spin-$\frac{1}{2}$ nuclei that interact in some manner can have their spins arranged either parallel or antiparallel only to each other. There are consequently two sorts of dihydrogen, H↑-H↑ that we call *ortho*-hydrogen (o-H_2) and H↑-H↓ which we call *para*-hydrogen (p-H_2). This has profound effects on the spectroscopy and thermodynamic properties of samples of dihydrogen since these normally contain both forms.

Under our usual assumption of separability of different contributions to the energy, the Schrödinger equation shows that the wave function can be written as the product of individual contributions (*cf.* the partition function):

$$\Psi_{total} = \varphi_{elec}\varphi_{vib}\varphi_{rot}\varphi_{ns} \tag{4.21}$$

The Pauli principle tells us that if a homonuclear molecule consists of spin-$\frac{1}{2}$ particles (called Fermions) this must change symmetry if the particles are interchanged. That is, the total wave function must be antisymmetric (*a*). But for spin-0 or integral spin nuclei (Bosons) there must be no such change and the wave function is symmetric (*s*).

Since the wave function can be written as a product we can assess the symmetry properties of each contribution under interchange and multiply them together using standard symmetry multiplication laws (*s.s* = *a.a* = *s*, *a.s* = *s.a* = *a*). For H_2 φ_{elec} is clearly symmetric since the molecular orbital is a $1s\sigma$ one and its sign does not change under inversion. Similarly φ_{vib} is symmetric for a molecule undergoing Simple Harmonic Oscillation in its lowest vibrational state. For H_2, therefore, we know that the Pauli principle requires that

$$\text{symmetry of } \Psi_{total} = s.s.\text{symmetry of } \varphi_{rot}.\text{symmetry of } \varphi_{ns} = a \quad (4.22)$$

since ^1H is a Fermion. Since *s.s* = *s* this implies that the product of nuclear spin and rotational symmetries can only be antisymmetric.

When the Schrödinger equation is solved for the rigid rotor model of the molecule we find that all wave functions for states with *J* even (0, 2, 4, . . .) have *s* symmetry, and all those with *J* odd (1, 3, 5, . . .) have *a* symmetry. So, to obey the Pauli principle, the nuclear spin wave function must have *a* symmetry when the rotational quantum number is even, and *s* symmetry otherwise (Figure 4.2). It remains to see what is meant by the nuclear spin wave function, and throughout we must remember the

		ϕ_{rot}	ϕ_{ns}	product
ortho	3 ————————	*a*	*s*	*a*
para	2 ————————	*s*	*a*	*a*
ortho	1 ————————	*a*	*s*	*a*
para	*J* = 0 ————————	*s*	*a*	*a*

Figure 4.2 *The lowest four rotational energy levels of H_2, showing the symmetries (symmetric and asymmetric) of the rotational and nuclear spin wave functions. The former are obtained from solution of the Schrödinger wave equation for the rigid rotor model of the molecule and they determine the symmetries required for the nuclear spin wave functions to make the product asymmetric, consistent with the Pauli principle. The energy levels with zero or even J are available to para-H_2 (this is the definition of para dihydrogen) and those with odd J are available to ortho-H_2. Unless a catalyst is present ortho–para interconversion is forbidden since it involves a change of symmetry of the nuclear spin wave function, and normal H_2 is a non-equilibrium mixture of the two forms.*

fundamental principle of physics that we cannot distinguish between identical particles.

We call a nucleus with spin ↑, α, and one with spin ↓, β, and we recall that there are just two relative orientations they can make with each other. Two possibilities, and two wave functions, are obvious, $\alpha(1)\alpha(2)$ and $\beta(1)\beta(2)$. That is the two spins on nuclei (1) and (2) may be parallel, both pointing either 'up' or 'down'. These are clearly two forms of o-H_2 and both have *s* symmetry since we cannot see any difference if we interchange the nuclei. It seems equally obvious that another form should be $\alpha(1)\beta(2)$, with the spins now antiparallel. But here we must be careful for we cannot tell which nucleus is which and it is equally likely that the spins are the other way round, $\alpha(2)\beta(1)$. A proper description must contain both possibilities, and we take care of this by taking linear combinations of them. But then quantum mechanics requires us to examine *all* linear combinations (and to introduce a normalisation factor of $1/\sqrt{2}$). The two wave functions are

$$\varphi_{ns} = \frac{1}{\sqrt{2}}[\alpha(1)\beta(2) + \alpha(2)\beta(1)] \qquad (4.23)$$

which is unchanged in sign if we interchange the nuclei and so is symmetric and another form of o-H_2, making three altogether, and

$$\varphi_{ns} = \frac{1}{\sqrt{2}}[\alpha(1)\beta(2) - \alpha(2)\beta(1)] \qquad (4.24)$$

This *does* change sign on interchanging the nuclei and has *a* symmetry. It is the nuclear spin wave function of p-H_2 and can be obtained in just one way.

To summarise, when ^1H–^1H occupies rotational levels of even *J* number it is in its *para*-form whereas when it occupies levels with odd *J* number it is in its *ortho*-form, Figure 4.2. Furthermore, there is a nuclear degeneracy of 1 associated with p-H_2 and 3 with o-H_2. The total degeneracy of the rotational/nuclear state is the product of this with $(2J + 1)$. Since many rotational levels are occupied at room temperature normal H_2 consists of mixtures of the two forms. However, interchange between the o- and p-forms proceeds at a detectable rate only in the presence of a catalyst and it is possible to separate them and perform experiments on either of the pure forms as well as on the normal mixture, and also on an equilibrium mixture produced in the presence of a catalyst. Their heat capacities all differ.

The rotational and nuclear spin partition function for the mixture that is normal H_2 has to be written as the sum of contributions from the two forms:

$$q_{rot,ns} = q_{rot,ns}(o\text{-}H_2) + q_{rot,ns}(p\text{-}H_2) = 3 \sum_{J\,odd} (2J + 1)e^{-BhcJ(J + 1)/kT}$$

$$+ 1 \sum_{J\,even} (2J + 1)e^{-BhcJ(J + 1)/kT} \qquad (4.25)$$

We could use this to calculate the thermodynamic properties of normal H_2, and the individual partition functions to obtain these for the two pure forms. Alternatively, we could do the same for an equilibrated mixture in which interchange between them is allowed. Some results for C_V are shown in Figure 4.3. The curve for pure $p\text{-}H_2$ exhibits a maximum due to rotational degeneracy effects (see Chapter 2) that does not appear in the pure $o\text{-}H_2$ curve, but addition of $3\times$ the latter to the former accurately reproduces the experimental curve for normal H_2 (this is not shown). Its form is vastly different from the curve calculated for equilibrated H_2.

To understand this latter observation we consider the relative concentrations of the two forms when equilibrium is established between them by adding a catalyst:

$$p\text{-}H_2 \leftrightarrow o\text{-}H_2$$

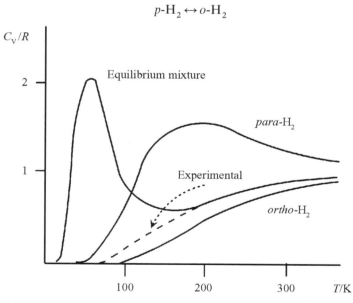

Figure 4.3 *Rotational/nuclear spin C_V of various forms of H_2. Experimental behaviour is reproduced by a 3:1 mixture of the heat capacities of pure* ortho *and pure* para H_2 *rather than by how an equilibrium mixture of the two forms would behave. This shows that, under normal conditions, no equilibrium exists and demonstrates the power of the symmetry arguments in the Pauli principle.*

for which the equilibrium constant is

$$K(T) = \frac{\text{Number of ortho molecules}}{\text{Number of para molecules}} = \frac{N_o}{N_p} \tag{4.26}$$

At thermal equilibrium the numbers are given by the Boltzmann equation and so, for example,

$$\frac{N_o}{N_o + N_p} = \frac{3\sum_{J_{odd}}(2J + 1)e^{-BhcJ(J + 1)/kT}}{q_{rot,ns}} \tag{4.27}$$

Consequently,

$$K(T) = \frac{3\sum_{J_{odd}}(2J + 1)e^{-BhcJ(J + 1)/kT}}{\sum_{J_{even}}(2J + 1)e^{-BhcJ(J + 1)/kT}} \tag{4.28}$$

At high temperatures (those where $T \gg \theta_{rot}$) so many rotational levels are occupied that to a very good approximation

$$\sum_{J_{even}}(2J + 1)e^{-BhcJ(J + 1)/kT} = \sum_{J_{odd}}(2J + 1)e^{-BhcJ(J + 1)/kT} \tag{4.29}$$

and $K(T) = 3$. Thus, at thermal equilibrium the gas consists of $\frac{3}{4}$ o-H_2 and $\frac{1}{4}$ p-H_2. This causes the Raman spectrum to exhibit an approximately $3:1$ ratio of intensities between adjacent lines since transitions from, *e.g.*, J_{odd} to $(J_{odd} + 2)$ originate in *ortho* molecules only whereas those between J_{even} and $(J_{even} + 2)$ originate in *para* ones.

When T is lowered to θ_{rot} the populations of the two forms in the equilibrated sample are about equal and as $T \to 0$ K, $N_o \to 0$ as does $K(T)$, all molecules being able to attain the lowest, *para*, level when the catalyst is present. The form of this variation in the composition of the mixture with temperature is what causes the change in C_V with temperature to differ so strongly from that of the normal, non-equilibrated, sample in which no molecules that start in the *para* state, for example, can attain the *ortho* one.

4.6 CHEMICAL EQUILIBRIA

We first establish the principles of our calculation by considering the simple and specific example of dissociation of a dihalogen molecule into atoms before generalising the results:

$$X_2 \leftrightarrow X + X.$$

The thermodynamic equilibrium constant is defined in terms of partial pressures as

$$K_P = \frac{p_X^2}{p_{X_2} p^\circ} \tag{4.30}$$

where p° is the standard pressure (1 bar).

From classical thermodynamics the Gibbs Free Energy change in the reaction is related to K_P through

$$\Delta G^\circ = -RT \ln K_P \tag{4.31}$$

where

$$\Delta G^\circ = 2G_X^\circ - G_{X_2}^\circ \tag{4.32}$$

We now enter our statistical thermodynamic expressions for these quantities, *e.g.*

$$(G^\circ - G(0)^\circ)_{X_2} = -RT \ln \frac{q_{X_2}^\circ}{N_A} \tag{4.33}$$

to obtain

$$\Delta G^\circ = [G(0)_{X_2}^\circ - 2G(0)_{X_2}^\circ] - RT \ln \left(\frac{q_m^\circ}{N_A} \right)_{X_2} + 2RT \ln \left(\frac{q_m^\circ}{N_A} \right)_X \tag{4.34}$$

where we have used the molar form of the partition functions. The superscript $^\circ$ throughout is to remind us that this must be written under the standard conditions of 1 bar to be consistent with the thermodynamic expression. As usual, we shall express the partition function as a product of independent factors, and this product contains the translational partition function, which is a function of the volume of the sample. But for a perfect gas the standard molar volume is obtained simply from $P^\circ V_m^\circ = RT$.

For a perfect gas at 0 K,

$$G(0) = H(0) - TS(0) = U(0) + PV - TS(0) = U(0) + nRT - TS(0) = U(0) \tag{4.35}$$

This means that the first term in Equation (4.34) is the molar internal energy difference (properly weighted for changes in stoichiometry) between the zero-point energy levels of reactants and products. This is given the symbol ΔE°. Inserting this and collecting up the remaining terms,

$$\Delta G^\circ = \Delta E^\circ + RT\ln\left(\frac{(q_m^\circ)_X^2}{(q_m^\circ)_{X_2}N_A}\right) = -RT\left\{\frac{-\Delta E^\circ}{RT} + \ln\left[\frac{(q_m^\circ)_X^2}{(q_m^\circ)_{X_2}N_A}\right]\right\} \qquad (4.36)$$

This can now be compared directly with the thermodynamic expression to give

$$\ln K_P = \frac{-\Delta E^\circ}{RT} + \ln\left[\frac{(q_m^\circ)_X^2}{(q_m^\circ)_{X_2}N_A}\right] \qquad (4.37)$$

or

$$K_P = \frac{(q_m^\circ)_X^2}{(q_m^\circ)_{X_2}N_A}\,e^{-\Delta E^\circ/RT} \qquad (4.38)$$

By inspection, this is straightforward to generalise for any equilibrium. Indeed the equilibrium between *ortho* and *para* hydrogen considered above could have been done in exactly this way to obtain the same result as given there. The pre-exponential term consists of the products of the partition functions of the products divided by that (in general those) of the reactant times the Avogadro number raised to the power of the number by which the number of product molecules differs from that of the reactant ones. This power is negative if the number of reactant molecules exceeds the number of product ones:

$$K_P = \frac{\Pi(q_m^\circ)_{\text{products}}}{\Pi(q_m^\circ)_{\text{reactants}}N_A^n}\,e^{-\Delta E^\circ/RT} \qquad (4.39)$$

where n = (number of products – number of reactants). [$\Pi(q)$ denotes the multiplication of the partition functions].

The exponential energy term is what we are long familiar with in chemistry, although defined with respect to zero-point energy differences. For dissociation of dihalogens it is simply the molar bond dissociation energy (D). It is *not*, however, the difference in heats of formation of the molecule and the atoms, as might be thought. This is because these quantities are defined at 298 K where many rotational levels are occupied, in contrast to just the lowest ones at 0 K.

We have now to enter our expressions for the partition functions. Atoms may only possess electronic and translational energy so for these,

$$q = q_{\text{elec}}q_{\text{trans}} \qquad (4.40)$$

We recall that the halogen atoms have np^1 electron configurations and so have two low-lying electronic states given by $^2P_{1/2}$ and $^2P_{3/2}$ (the lower), separated by ε in energy, with degeneracies of 2 and 4 respectively, giving

$$q_{\text{elec}} = 4 + 2e^{-\varepsilon/kT} \tag{4.41}$$

This actually has to be written with caution since the dissociation may produce atoms in one of these two electronic states only (see Problem 4.5).

The translational term is

$$q_{\text{trans}} = \left\{\frac{2\pi m_X kT}{h^2}\right\}^{3/2} V_m^\circ = \left\{\frac{2\pi m_X kT}{h^2}\right\}^{3/2} \frac{RT}{P^\circ} \tag{4.42}$$

so that overall for the atom,

$$q_X = (4 + 2e^{-\varepsilon/kT})\left\{\frac{2\pi m_X kT}{h^2}\right\}^{3/2} \frac{RT}{P^\circ} \tag{4.43}$$

The molecule has no unpaired electron and therefore no spin–orbit coupling so that the electronic ground state has a degeneracy of one and $q_{\text{elec}} = 1$. However, it may have non-unity vibrational, rotational and translational contributions

$$q_{X_2} = q_{\text{elec}} q_{\text{vib}} q_{\text{rot}} q_{\text{trans}} = 1\left(\frac{1}{1 - e^{-h\nu/kT}}\right)\left(\frac{kT}{\sigma hcB}\right)\left(\frac{2\pi m_{X_2} kT}{h^2}\right)^{3/2} \frac{RT}{P^\circ} \tag{4.44}$$

where we have taken the lowest vibrational level to be of zero energy and have substituted for the standard molar volume. For the homonuclear diatomic dihalogen molecule $\sigma = 2$.

These expressions are now inserted into that for K_P to give

$$K_P = \frac{\left[\left(4 + 2e^{-\varepsilon/kT}\right)\left\{\frac{2\pi m_X kT}{h^2}\right\}^{3/2} \frac{RT}{P^\circ}\right]^2 e^{-D/RT}}{\left(\frac{1}{1 - e^{-h\nu/kT}}\right)\left(\frac{kT}{2hcB}\right)\left\{\frac{2\pi m_{X_2} kT}{h^2}\right\}^{3/2} \frac{RT}{P^\circ} N_A} \tag{4.45}$$

This could be simplified but we shall leave it in this form, which shows how the equation is obtained. The expression looks formidable but it has been a straightforward business of substituting our previous expressions for partition functions into our expression for K_P to get it, and is quite automatic.

At this stage the power of the statistical thermodynamic approach is evident. We would have known from classical thermodynamics to expect the exponential term. But we would not have guessed that the pre-exponential one was determined by such properties as the vibrational frequency of the bond, the rotational constant, the masses of the atoms and molecules and, possibly, the existence of spin–orbit coupling in the atoms. The masses of the atoms also directly affect the values of the vibrational frequency and the rotational constant, and the analysis gives a clear picture of why the pre-exponential factors for different dihalogen molecules differ.

4.6.1 Isotope Equilibria

An interesting case of equilibrium occurs in the exchange reaction between the isotopic forms of the dihalogens, for example

$$^{35}Cl-^{35}Cl+ {}^{37}Cl-^{37}Cl \longleftrightarrow 2{}^{35}Cl-^{37}Cl$$

Here the number of reactant and product molecules is the same so that no term in N_A occurs in the expression for the equilibrium constant $[(N_A)° = 1]$, and the terms in $RT/P°$ cancel. We make the further simplifying assumption that $q_{vib} = 1$ for all species (this is not a necessary assumption but it shortens the calculation). Furthermore, the bond energy of all the species is the same so that $\Delta E° = 0$ and the exponential term in the energy has a value of 1.

It follows that

$$K_P = \frac{(q_{rot}q_{trans})^2_{35,37}}{(q_{rot}q_{trans})_{35,35}(q_{rot}q_{trans})_{37,37}} = \frac{q^2_{rot,35,37}}{q_{rot,35,35}q_{rot,37,37}} \frac{q^2_{trans,35,37}}{q_{trans,35,35}q_{trans,37,37}} \quad (4.46)$$

Entering the masses of the molecules into the partition functions shows that the translational energy factor $\sim 1.008 \approx 1$ so that only the rotational term remains significant. We now insert the expressions for the partition functions:

$$K_P = \frac{\left(\frac{kT}{\sigma hcB}\right)^2_{35,37}}{\left(\frac{kT}{\sigma hcB}\right)_{35,35}\left(\frac{kT}{\sigma hcB}\right)_{37,37}} \quad (4.47)$$

The isotopomers have identical bond lengths, and entering the atomic masses into the calculation of the rotational constants, B, shows that their product ratio is again very close to 1, with the result that to a very good approximation

$$K_P = \frac{\sigma_{35,35}\sigma_{37,37}}{\sigma_{35,37}^2} = \frac{2.2}{1^2} = 4 \qquad (4.48)$$

The remarkable fact, therefore, is that the equilibrium constant is almost entirely determined by the symmetry numbers of the various iso-topomers. This is a telling illustration that apparently obscure fundamental properties (here the Pauli principle) can affect real chemistry. The significance of this value would not be at all evident without the statistical mechanical approach but would rather remain an unexplained experimental observation.

4.7 CHEMICAL REACTION

It might seem strange that in a chapter devoted to thermodynamics we now turn to kinetics but we do so in an attempt to obtain the same depth of understanding of the molecular properties that determine reaction rates. We start from the best-known characteristic of reaction rate constants: the great majority of them vary with temperature in a way described in the Arrhenius equation

$$k_r = Ae^{-E_A/RT} \qquad (4.49)$$

where E_A is the activation energy of the reaction. The pre-exponential factor A appears as a temperature-independent term whereas careful experiment shows that it does in fact have a temperature dependence which differs with the complexity of the specific reaction involved. We should like to understand this and predict the rate constant at any given temperature.

The approach we take is through Transition State (TS) theory, also known as Activated State theory (Figure 4.4). In this, reaction proceeds by supplying sufficient energy for the reactant (we consider a dihalogen, as before) to overcome an activation energy barrier and proceed to the product (halogen atoms). The configuration of atoms at the top of the energy barrier is called the Transition State. In recent years it has been directly observed experimentally in some reactions using very fast observational methods (necessitated by its short lifetime) based upon pulsed lasers whose light flashes are of a few femtosecond (10^{-15} s) duration. The basic assumption (perhaps best justified by finding that it leads to predictions consistent with experiment) is then made that this is in equilibrium with the reactants, but that some of the system continually leaks over the barrier to form products. The rate at which it does so clearly depends on the concentration of the TS:

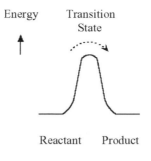

Figure 4.4 *Schematic drawing of the progress of a reaction from reactant to product. The reaction co-ordinate is the lowest energy path between the two, and that preferred by the reaction; in detail it is a cut across the total potential energy surface of the reaction (Chapter 5). Reactant and product are linked through a maximum in the potential energy that must be crossed for reaction to occur. It corresponds to a distinct structure and is termed the transition state. The height of the maximum relative to the lowest energy in the reactant represents the activation energy for the forward reaction.*

$$X_2 \xleftarrow{K_C^{TS}} TS \xrightarrow{k} 2X \qquad (4.50)$$

Here we have been careful to remind ourselves that the equilibrium constant must be defined with respect to concentration since the rate of a reaction is *concentration* dependent. If the reaction is found experimentally to be first-order with rate constant k_{obs}, using this equation allows us to write

$$\frac{dX}{dt} = k_{obs}[X_2] = k[TS] \qquad (4.51)$$

where the fact that two atoms are produced in each dissociation has been included in k_{obs}. But since the TS is in equilibrium with the reactant,

$$K_C^{TS} = \frac{[TS]}{[X_2]} \text{ or } [TS] = K_C^{TS}[X_2] \qquad (4.52)$$

Combining the two equations gives

$$k_{obs} = k K_C^{TS} \qquad (4.53)$$

Thus far we have used no statistical thermodynamics at all and this result is a general one of TS theory that can be evaluated in several ways. We might, for example, choose to put in a thermodynamic expression for K_C in terms of the Gibbs Free Energy of activation. Here, however, we seek molecular insight and so turn to statistical thermodynamics. But we have seen that this leads most directly to an expression for K_P and we have first to express K_C in terms of it. A standard thermodynamic result tells us that for 1 mole of a perfect gas under standard conditions

$$K_C^{TS} = \left(\frac{RT}{P^\circ}\right) K_P^{TS} \tag{4.54}$$

From Section 4.6 we have seen that K_P contains a term in the energy change between reactants and products at 0 K. Here the equilibrium is between reactants and TS and the change, the activation energy, is given by

$$E_A = E^\circ(TS) - E^\circ(X_2) \tag{4.55}$$

Substituting these relationships and the expression for K_P from the last section gives

$$k_{obs} = kK_P^{TS}\frac{RT}{P^\circ} = k\frac{RT q_{m,TS}^\circ}{P^\circ q_{m,X_2}^\circ}e^{-E_A/RT} \tag{4.56}$$

It remains to express the partition functions as products of individual terms and then to substitute our expressions for them. In this example X_2 is a diatomic, and so evidently must also be the TS. We are dealing with both in their singly-degenerate electronic ground states so that for both $q_{elec} = 1$, leaving the vibrational, rotational and translational contributions to be considered. This requires some thought for the value of q_{vib} of the TS since a diatomic possesses only one vibrational degree of freedom through having just one bond, but this bond breaks in the reaction. For this vibration the frequency (v_{TS}) must be extremely low since one vibration of sufficient energy is enough to break the bond, whereas if no bond is broken a typical vibration frequency is $\sim 10^{13}$ Hz. We can therefore expand the exponential, truncate it after a single term, and still make a very good approximation for q_{vib}:

$$q_{vib,TS} = \frac{1}{1 - e^{-hv_{TS}/kT}} = \frac{1}{1 - \left(1 - \frac{hv_{TS}}{kT} + \cdots\right)} \approx \frac{kT}{hv_{TS}} \tag{4.57}$$

Consequently,

$$q_{m,TS}^\circ = \left(\frac{kT}{hv_{TS}}\right)q_{rot,TS}^\circ q_{trans,TS}^\circ \tag{4.58}$$

We now consider the reaction step and the meaning of the rate constant k

(the use of the same symbol for the rate constant and the Boltzmann constant is another unfortunate historical accident). From the above it might be thought that every time that the transition state undergoes a vibration it dissociates into atoms, and the rate constant would simply be v_{TS} s^{-1}. But this is not the case since we could not set up the equilibrium between reactant and TS if it was true. We therefore introduce a factor, the 'transmission coefficient, κ', to account for this:

$$k = \kappa v_{TS} \tag{4.59}$$

Inserting these equations into that for k_{obs} allows v_{TS} to be cancelled and gives

$$k_{obs} = \kappa \left(\frac{kT}{h}\right)\left(\frac{RT}{P^\circ}\right)\frac{q^\circ_{rot,TS}q^\circ_{trans,TS}}{q^\circ_{vib,X_2}\, q^\circ_{rot,X_2}q^\circ_{trans,X_2}}\, e^{-E_A/RT} \tag{4.60}$$

All qs remain molar functions. This expression clearly has the same form as the Arrhenius equation and it is interesting to note how straightforwardly the exponential term has arisen. But now we have an expression for the pre-exponential factor A and can evaluate it simply by substituting our expressions for the individual partition functions as was done in the equilibrium constant calculation above. Therefore, A depends on similar physical properties of the reactant and the TS. When this is done it produces very satisfactory agreement with experimental results for reactions of varying complexity.

We return, however, to our initial problem, that of the temperature dependence of A. There is an overt dependence on T^2 in our expression but we must remember that the individual contributions to q have their own temperature dependences that we now substitute in; q_{rot} depends on T, q_{trans} on $T^{3/2}$. The dissociation of at least the heavier of the dihalogens occurs at sufficiently low temperatures that $q_{vib} \sim 1$ and is temperature independent.

$$A \propto T^2\frac{T.T^{3/2}}{T.T^{3/2}} = T^2 \tag{4.61}$$

Had we calculated the rate constant of the reverse reaction,

$$X + X \longleftrightarrow TS \longrightarrow X_2$$

we would have obtained

$$k_{obs} = \kappa \left(\frac{kT}{h}\right)\left(\frac{RT}{P^\circ}\right)\frac{q^\circ_{rot,TS}q^\circ_{trans,TS}}{(q^\circ_{elec,X}q^\circ_{trans,X})^2 N_A^{-1}}\, e^{-E_A/RT} \tag{4.62}$$

where the activation energy is in general different from that above. Note that the structure of the equation is that the partition function of the TS is still the numerator, and the product of those of the reactants is in the denominator. Also, since the number of molecules diminishes by 1, the factor of N_A^{-1} appears in the denominator. We have remembered, of course, to include q_{elec} for the halogen atoms. Now the temperature dependence (neglecting, as is usually possible, changes to q_{elec}) is

$$A \propto T^2 \frac{T.T^{3/2}}{(T^{3/2})^2} = T^{3/2} \tag{4.63}$$

This is different from that above. It is a general result that the form of the pre-exponential factor, and its temperature dependence, varies with the number of molecules involved in the reaction.

The treatment is easily extended to more complicated reactions. In a reaction between two diatomic molecules, for example, the TS would contain 4 atoms with (unless it happened to be linear when there would be one more) $(3 \times 4 - 6) = 6$ vibrational degrees of freedom. With the same treatment of the breaking bond as above we obtain the factor (kT/h) but there remain 5 vibrational contributions and 3 rotational ones for the non-linear molecule, along with the translational contribution. If the reaction proceeds at low enough temperature, all 5 of these vibrational partition functions may be approximately 1, depending on the precise molecules involved. Otherwise, we must assume a structure for the TS and use vibrational frequencies measured using analogous stable molecules. A similar problem is found for rotation in which a structure again has to be assumed. The denominator would contain the full partition functions of the two reactants, 6 factors in all (provided $q_{elec} = 1$ for each molecule). But no new principles are involved, and whilst the problem appears complicated the equation is perfectly straightforward to write down (Problems 4.6 and 4.7).

The statistical thermodynamic version of TS theory is remarkably successful in predicting rate constants in agreement with experiment, no matter how complex the reaction is and despite the number of assumptions made.

4.8 THERMAL EQUILIBRIUM AND TEMPERATURE

Throughout this book we have stressed that we have been dealing with the ensemble properties of systems that are at thermal equilibrium with their surroundings, and have warned that the results may not be applicable to modern experiments performed on a small number of molecules, or even a single one. However, there are circumstances where they can be and a case in which this can be seen is in the concept of temperature.

Temperature is, classically, an ensemble property that can be defined in terms of the Maxwell distribution of velocities. In a sample even of as few as a hundred atoms, say, we may measure this distribution experimentally (using velocity selectors, Chapter 5) and fit it to the Maxwell equation (Equation 1.4) to derive a value for the temperature. This means that we do not need a large ensemble for temperature to be a valid concept. But some experiments are now conducted on single molecules that form a part of an ensemble and measurement of a velocity distribution becomes impractical. However, we can make recourse to the principle that the time-average behaviour of a system is the same as its ensemble average. This implies that if the rotational populations of a molecule, say, are monitored over infinite time the result is the same as if an ensemble of identical molecules are monitored instantaneously. We can therefore obtain a value of the temperature by fitting the populations to the Boltzmann distribution.

The validity of this approach depends upon the molecule being able to come to thermal equilibrium in the sense described in Section 1.6. That is, following absorption of energy the molecule attains a higher energy level from which it rapidly loses energy to return to the ground state before absorbing energy once again, in general to attain a different higher level than before. This has implications as to what we mean by 'infinite time', for the molecule must explore all the energy states available to it at the temperature of the ensemble before the approach is valid. In a gas at atmospheric pressure the molecule makes very frequent collisions with others (Section 1.2) and on each collision is able to lose energy to them so that the relaxation process to the lower levels is very efficient. If we monitor just one molecule inside an ensemble even over a finite period of time we can therefore measure its temperature; this is possible in modern experiments. However, an isolated molecule in a vacuum cannot attain equilibrium with its surroundings since it does not have any! It does not have a temperature in the normal sense. But if in an experiment an isolated atom in its electronic ground state is kept perfectly stationary in space we would know that its temperature was 0 K. This has almost been attained in practice by keeping the atom in place by collisions with photons from suitably directed laser beams.

Experimentally, we often define the temperature of a system in terms of the populations of energy levels, and since a molecule may possess electronic, vibrational and rotational energy there are three associated 'temperatures'. At room temperature, only the electronic ground state of most molecules is occupied so that, using the Boltzmann distribution, we say that the electronic temperature is 0 K. The same may be true of the vibrational temperature. But we have seen that many of the rotational levels associated with the electronic and vibrational ground states are occupied, and so the sample has a non-zero rotational temperature. A translational temperature can be defined in terms of the motion of the molecules. These four temperatures often differ. An example is found in molecular beam experiments in which a fast-moving beam of molecules is made by expanding a gas from a high pressure source into a vacuum under 'supersonic nozzle' conditions. The gas may have a high translational temperature but nevertheless usually has most of the rotational population in the lowest level, and therefore a low rotational temperature. However, in an ensemble at equilibrium at temperatures well above θ_{rot} the rotational and translational temperatures are equal and either measures what we normally understand as the 'temperature' of the system. In practice we tend to use these ideas in the opposite sense: if molecules possess, for example, a vibrational temperature other than 0 K it tells us that some are in a higher vibrational level.

PROBLEMS

Any constants required are listed at the end of the book.

4.1   The entropies of the rare gases at 298 K are He, 126.0; Ne, 146.2; Ar, 154.7; Kr, 164.0; and Xe, 170.0 J K^{-1} mol^{-1}. Their relative atomic masses of samples containing normal isotopic mixtures are 4, 20.2, 39.95, 83.8 and 131.8. Use the Sackur–Tetrode equation (Equation 4.4) to show that these values are predicted by the theory.
   (Hint: substitute RT/P° for V_M°, the standard molar volume.)

4.2   Ne and HF have the same molecular mass and yet their molar entropies at 298 K are 146.2 and 174 J K^{-1} respectively. Explain why these differ and rationalise the value for HF.
   (The rotational constant for HF is 21.95 cm^{-1}.)
   H_2O also has the same mass and yet its molar entropy is 189 J K^{-1}. Comment.

4.3   Calculate the molar bulk susceptibility, χ, of a system containing free radicals at 298 K, and calculate the total magnetisation, M, in a field of 10 T.
   Inside a magnetic field of strength B the energy of such a sample is MB. What is the magnetic energy of a molar sample in a field of 10 T? How high would the field have to be to make the energy comparable with RT at 298 K?

4.4   In the infrared spectrum of ethyne ($^1HC \equiv C^1H$) the rotational transitions associated with the asymmetric stretch vibration show intensity variations in the ratio 3:1. Discuss.

(Ethyne, like $^1H_2$, is a symmetric linear molecule.)

4.5   Calculate the equilibrium constant, K_P, at a constant pressure of $P°$ and 500 K for the dissociation $I_2 \longleftrightarrow I + I$.

This reaction is more complicated than it would appear since the I atoms are formed in the $^2P_{3/2}$ state only, so that the electronic contribution to the partition function is 4 *not* $(4 + 2e^{-\varepsilon/kT})$.

For this molecule, $B = 0.037 \ cm^{-1}$, $hv = 214.5 \ cm^{-1}$ and the bond dissociation energy $D = 12\,453 \ cm^{-1}$. The first electronic state of the I_2 molecule lies $8124 \ cm^{-1}$ above the ground state and, together with all other electronic states, can be ignored.

4.6   (i) Using Transition State theory obtain an algebraic expression for the rate constant for a reaction involving two atoms of mass m and n coming together to form a di-atomic transition state (and a diatomic product).

(ii) Since translational partition functions are very large they are to a first approximation, independent of the mass of the atoms. Make this assumption and simplify the expression by setting all equal.

4.7   Repeat the calculation in Problem 4.6 making the same assumption as in 4.6(ii) for a generalised reaction involving polyatomic reactants with internal modes. With these there are so many energy levels that all rotational contributions to the partition function from different molecules can be taken to be approximately equal, and all vibrational contributions are also equal if, as we assume, all vibrational modes are excited. A non-linear molecule with N atoms possesses $3N - 6$ vibrational modes, 3 rotational ones and 3 translational ones. If the second reactant possesses N^* atoms, the transition state contains $(N + N^*)$ atoms and undergoes $3(N + N^*) - 6$ vibrations, of which one vanishes in the reaction to give $3(N + N^*) - 7$ vibrations to be considered, along with the usual 3 for rotation and 3 for translation. Due to their equality the contributions from the internal modes of the various species can be cancelled.

By comparing the results of the calculations with that of problem 4.6(ii) show that the ratio of the pre-exponential factors of reactions between atoms and reactions involving complex molecules is approximately $(q_{vib}/q_{rot})^5$.

Evaluate this for a typical ratio for large molecules of 50.

This important approximate calculation predicts that for two reactions occurring with the same activation energy that involving complex molecules is about 10^9 times slower than one involving atoms (see Chapter 5).

Reactions

5.1 INTRODUCTION

In Chapter 4 we saw an approach to calculating the rate constant of a chemical reaction using Transition State theory, and pointed out that there is direct experimental evidence that the Transition State exists. But the theory was based upon the statistical thermodynamic approach we have developed and we deduced the properties that affect reaction rates through this. Although this has arisen using a logical development of thermodynamics through the properties of atoms and molecules this is an indirect way of finding out how reactions actually occur and in a sense is a denigration of the whole principle of this book. How much better it would be if we could observe directly what happens when one molecule encounters another in a reaction and build up our understanding of reactions from this. This is now possible and in this chapter we indicate the principles involved without becoming concerned with the details that would need a whole book devoted to chemical kinetics.

5.2 MOLECULAR COLLISIONS

Evidently, for two molecules to react they must in some sense encounter each other. If they were hard spheres they would have to collide. Any contact might do, ranging from a full-on collision to a grazing one. That is, a collision is said to occur if the spheres touch in any sense. A geometrically equivalent but more convenient way of expressing this is that their centres should pass within the sum of their radii. For two identical molecules this implies a collision cross-section, $\sigma_{\text{collision}} = \pi d^2$, where d is twice the radius. This is actually the cross-sectional area of each molecule. For non-identical molecules the collision diameter is the sum of the radii of the individual species. But real molecules are not hard spheres and we must investigate how close one molecule must approach another *for reaction to occur*, acknowledging that the molecules might not have to collide in the solid sphere sense. This leads to the concept of a reaction cross-section, $\sigma_{reaction}$, which in general may differ from $\sigma_{collision}$. It is only at our most simplistic that we think of a molecule having a definite (hard sphere) diameter and indeed there are several definitions of molecular diameter according to what property of the molecule is involved; we have introduced both collision and reaction cross-sections but a third definition might, for instance, be in terms of the extent of the electron distribution.

5.2.1 Collision Diameters

We need first to understand the meaning and origin of the collision cross-section. For this we consider how the energy of interaction between molecules changes as one molecule (we shall in fact concentrate on atoms) approaches another. This is shown in the potential energy diagram (Figure 5.1) for a pair of neon atoms. All atoms and molecules attract each other through intermolecular forces as they approach (in this case

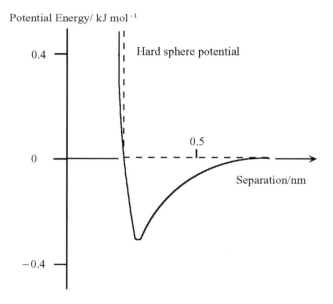

Figure 5.1 *Potential energy curves for neon atoms (——) and for hard spheres (– – –). Apart from the shallow minimum the neon curve is quite well approximated by the hard sphere one, with the repulsive part rising rapidly as the inter-nuclear separation is decreased. A slow-moving incoming atom is repelled once it reaches the cross-over point in the curve but if it is speeded up repulsion does not occur until the repulsive energy exceeds the kinetic energy of the approaching atom. In this way the distance of closest approach can be determined over a range of kinetic energies, and the repulsive part of the curve plotted.*

through London forces, which share with all types of attractive inter-molecular forces observed in molecules a distance dependence on r^{-6} where r is the inter-nuclear separation) but then repel each other as their nuclei become sufficiently close for their electrostatic interactions to dominate. This leads to a characteristic form of the potential energy curve. Looking at Figure 5.1 it might be thought that neon should exist as Ne_2, for it is apparent that there is a minimum in the curve, implying that the 'molecule' with a separation equal to where this minimum occurs is more stable than two isolated atoms. But the minimum is very shallow in energy, and the di-atomic molecule is simply blown apart every time it collides with another atom, which we saw in Chapter 1 occurs very frequently. At very low temperatures, where kT is less than the attractive energy, diatomic molecules of the rare gases can be observed experimentally. They are known as 'Van der Waals' molecules, reflecting the origin of the weak, non-chemical binding, forces that hold them together.

The figure shows that as one atom approaches the other the energy of the system is lower than that of the two separate atoms until the point is reached where the curve crosses the energy axis where there is a net repulsion. This, therefore, corresponds to the collision diameter since the atoms get no closer. But then the energy rises very rapidly with the distance of approach (typically as r^{-14}) and, together with the minimum being very shallow, the whole curve is quite well approximated by the hard sphere potential. Inert gases do behave largely as though they are hard spheres.

The most direct method for measuring the collision cross-section is to use a molecular beam experiment. A beam of atoms or molecules is created in which they move at constant velocity without colliding with each other in the beam. One way of making such a beam is to use a low pressure gas source in which there is a Maxwellian distribution of velocities determined by its temperature. If the pressure in the source is sufficiently low, molecules effuse from this region into a vacuum region where they undergo velocity selection to ensure that only those with a certain velocity pass into the experimental region. This is done mechanically. The beam passes through a series of toothed disks rotating at a constant speed. A molecule that passes through a gap in the first only passes through one on the second if it travels between the two at such a rate that a gap in the second is in the right place to allow this. Otherwise it strikes the second disk and is pumped away. This is most easily understood by considering two disks, each with a single hole in its periphery, mounted on a solid rotating axle so that the holes are not opposite (Figure 5.2). Increasing the number of holes in a toothed wheel merely increases the intensity of the beam. After several stages of velocity selec-

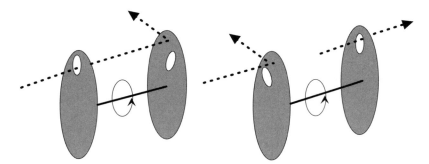

Figure 5.2 *Principle of velocity selection. In this simplest case two solid discs each containing a single hole are attached to a rigid axle that is rotated. An approaching
beam of molecules may enter the region between the discs if it gets to the first hole
at the right time. It then normally strikes the second disc, bounces off and is lost.
However, if during the transit time between the two discs the second has rotated
to present its hole to the incoming beam then it gets through this, too, although
molecules which strike the first disc at this time are now lost. By adjusting the
speed of rotation of the disc any velocity can be selected from the Maxwellian
distribution that exists when the molecules leave the low-pressure hot source.*

tion, involving a series of discs mounted on the same shaft, a very narrow
velocity band can be selected. A second method, which produces a much
more intense beam, is to use the hydrodynamic properties of the so-called
'supersonic nozzle'. This consists of a high pressure gas source feeding
directly into the vacuum through a specially-shaped nozzle, rather like a
pea-shooter.

To measure the collision cross-section the neon beam (say) impinges on
a target gas of neon atoms from which the atoms in the beam bounce off.
They are said to be 'scattered' and from an analysis of their scattering
directions and intensities based upon the classical laws of conservation of
energy and momentum the collision cross-section can be assessed. Previous to this experiment the cross-section was deduced from measurements
of the viscosity of gases using the kinetic theory of gases, based upon
equations derived from the Maxwell distribution. But the beam experiment allows a further step. By increasing the speed of the atoms in the
beam their kinetic energy is increased, and now the molecules approach
until they are repelled when the repulsion energy exceeds this. Use of
faster and faster beams of known energy allows the variation of the
collision cross-section with energy to be obtained, and hence the dis-
tance-dependence of the repulsive part of the potential energy curve can
be investigated directly.

An important concept in collisions is that of the 'impact parameter, b',
which is a measure of the extent of a collision. It is the perpendicular

distance between the centre of the atom or molecule hit in the collision and the linear direction of the approaching molecule. For hard sphere collisions impacts occur between $b = 0$ (a direct hit) and $b = d$ (the grazing collision).

Knowledge of the collision diameter for a particular encounter provides a datum with which a measured reaction cross-section can be compared.

5.2.2 Reactive Collisions

To study reactive collisions a variation on the beam method is employed. Now we make two molecular beams, one from each of the reactants, and collide them before determining the directions in space in which the products fly out, using a moveable detector (Figure 5.3), often a mass spectrometer. This experiment is technically difficult and is restricted to reactions in which sufficiently intense beams can be made. Analysis of the results is also demanding. It is again performed using the classical laws of physics but in a co-ordinate system based upon the centre of mass of the reactants. Here we shall ignore such complications (and, largely, the velocity information that can be extracted) and simply depict in our figures the directions in which the products are ejected from the reaction. The resulting diagrams are simplifications, but they contain the information important to the chemist in seeing what happens in a reaction.

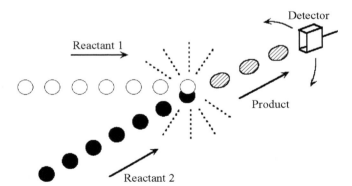

Figure 5.3 *In each of two molecular beams the molecules travel at constant speed (which may differ in the two beams), and do not therefore collide with each other. However, in the crossed beam experiment the two beams are brought into collision, the molecules react and products are ejected along specific directions in space. They are observed using a detector that is moved on the surface of a sphere centred at the position of the collision. From measurement of the angular dependence of product formation the geometry of the actual reaction process can be worked out.*

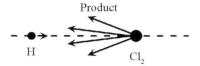

Figure 5.4 *When H atoms approach a Cl_2 molecule and react the product is backward scattered along the direction of the incoming atoms. Reaction occurs only over small values of the impact parameter, collision having to be essentially 'head on'. For this reaction the reaction cross-section is to a good approximation equal to the collision cross section. In all similar diagrams it should be remembered that the experiment is in three dimensions; what is shown as a fan of products here is a projection of an actual cone.*

The first example we consider is the reaction of H atoms with Cl_2 molecules (Figure 5.4). Here the HCl product is 'backward scattered'. That is it is thrown out against the direction of approach of the H atom to the molecule. It is quite like a hard sphere collision except it is HCl that comes off rather than the H atom rebounding. The directions in which the product is ejected are cylindrically symmetric about the direction of approach. These two features are characteristic of a collision process and there is no direct evidence for a transition state. However, we saw in the previous chapter that this is very short lived, and its existence would not be shown in this experiment. Analysis shows that for this reaction the reaction cross-section is equal to the collision one, implying that for a reaction to occur the reactants must become extremely close.

Now we consider the reaction of K atoms with Br_2 molecules, which exhibits a very different scattering pattern (Figure 5.5). Now some backward scattering of product is seen as before but there is a more substantial forward scattering too, and it is immediately obvious that the reaction cross-section is greater. The backward scattering is again due to collisions with low impact parameter, but the forward scattering implies that the reactants do not have to collide in the crude physical sense to yield

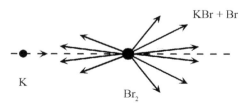

Figure 5.5 *When a K atom reacts with a Br_2 molecule some of the product is back-scattered but rather more is forward-scattered along the direction in which the atom approaches. The process has the characteristics of a collision but the reaction cross-section is increased compared with that in Figure 5.4 and is greater than the collision cross-section. The forward scattering occurs via the harpoon mechanism.*

products. Rather it shows that the reaction can also occur with an incoming trajectory with impact parameter $> d$, that is it happens between comparatively widely separated molecules. Note, however, that the cones of the product distribution are about the direction of approach, as before, and the reaction has the characteristics of a collision. There is again no indication of a transition state sufficiently long-lived to affect this distribution.

So what is the origin of the forward-scattered product? The reaction is an example of a 'harpoon reaction'. As the K atom approaches the very electronegative Br_2 molecule an electron jumps between them to form two ionic species, K^+ and Br_2^- which then attract each other electrostatically according to an inverse square law in their distance of separation. This is much longer-ranged in its action than the r^{-6} law of attractive intermolecular forces and reels the reactants in to close proximity so that reaction occurs to give KBr and Br atoms. The net result of the early electron transfer is to increase the reaction cross-section to 5.5 times the collision cross-section. The reaction provides a very clear illustration of the crucial difference between the two.

Not all reactions studied in molecular beam experiments show the simple characteristics of collisions however, in that their product distribution is spherically symmetric and no longer bunched along the direction of approach. An example is the reaction of Cs atoms with RbCl molecules (Figure 5.6). This pattern occurs because a tri-atomic intermediate is formed of sufficient lifetime that it undergoes rotation before dissociating into the products, CsCl and Rb atoms. The products are thrown out as the rotation occurs, in a similar way to how a catherine wheel firework throws out sparks as it rotates. The lifetime of the intermediate, which we might be tempted to recognise as the transition state, is therefore of the order of a rotation period, 10^{-10} s. However, with an intermediate this long-lived it should be thought of as a metastable species separated from both reactants and products by two true transition states, one involved with its formation and the other with its dissociation. That is it exists in a shallow local potential energy minimum separated by maxima on both the reactant and product sides. It is obviously more difficult in this case to obtain a value for the reaction cross-section from the results. Nevertheless a simple ruse allows us to encompass it within the collision theory of chemical reactions (see below).

Finally, many chemical reactions occur most efficiently, or maybe only occur at all, if the reactants approach each other along definite directions in space, a steric effect. This is well known in the S_N1 and S_N2 reaction mechanisms of organic chemistry. It, too, can be demonstrated directly in a molecular beam experiment with, for example, the reaction cross-

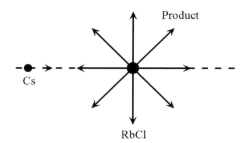

Figure 5.6 *In contrast to the previous cases, when Cs atoms react with RbCl molecules the product is uniformly distributed over a sphere and there is no tendency for it to group about the direction of approach. The spherical distribution suggests that a tri-atomic complex is formed that persists long enough to rotate and throw off products with equal probability in three dimensions.*

section for the $K + CH_3I$ reaction being greatest if the atom approaches the iodine end of the molecule rather than the methyl group. This ingenious experiment requires orienting the methyl iodide molecule in space by a combination of strategically placed electric fields and electric field gradients. But the product distribution characteristics are those of a collision whichever side the atom approaches the molecule. If the reaction is performed in a normal laboratory rather than with crossed molecular beams the methyl iodide molecules are randomly oriented with respect to the incoming atom and an average cross-section can be deduced from measurement of the reaction rate constant (see below).

Before leaving this section we stress that only the crudest interpretation of the results has been given, and more detail can be obtained from a fuller analysis. The experiments provide direct evidence of what happens in the collisions of molecules that lead to chemical reaction.

5.3 COLLISION THEORY

The simplest theory of chemical reaction rates, the collision theory, calculates the number of collisions that occur with sufficient energy for reaction to occur (the activation energy E_A) and equates this to the reaction rate. This leads to an expression for the rate constant of a bimolecular reaction (usually expressed in $dm^3\ mol^{-1}\ s^{-1}$)

$$k = Z_{AB}^{reaction}e^{-E_A/kT} = \sigma_{reaction}\left(\frac{8kT}{\pi\mu}\right)^{\frac{1}{2}}N_Ae^{-E_A/kT} \qquad (5.1)$$

where μ is the reduced mass of the molecules or atoms involved. The pre-exponential factor gives the absolute number of collisions ($Z_{AB}^{reaction}$) at the reactive distance per unit volume per unit time, divided by N_A to obtain the result in molar units, whilst the exponential term gives the fraction of collisions that occur with energy $> E_A$ (from the Boltzmann distribution in one dimension). The former is obtained from the kinetic theory of gases, found in any physical chemistry textbook, and we have simply replaced the collision cross-section appropriate to hard sphere collisions by the reaction cross-section relevant to reactions. These, as we have seen, are not necessarily the same but we can relate them through the empirical expression

$$\sigma_{\text{reaction}} = P\sigma_{\text{collision}} \tag{5.2}$$

Insertion of this into Equation (5.1) yields the expression for the rate constant normally quoted in collision theory. P is known as the steric factor, and direction of approach and steric hindrance are indeed major factors in determining its value in most cases. Defined in this empirical way we can use it to encompass all the situations we have met above, including the $K + Br_2$ reaction and even the $Cs + RbCl$ one. It is 5.5 for the former but only $\sim 10^{-6}$ for the hydrogenation of ethene. A limiting value for reactions involving molecules containing many atoms appears to be $\sim 10^{-9}$. Conventionally, rate constants are discussed in terms of the P factor, but it is apparent from above that we might more meaningfully use the experimental results to determine the reaction cross-sections directly. This provides a method for direct comparison between normal laboratory and molecular beam results.

It is rather unsatisfactory that the calculation using collision diameters may overestimate the rate of reaction one-billion-fold. We should remember, however, that Transition State theory provides a means for calculating the pre-exponential factor in the Arrhenius equation and one of its triumphs is that for reactions involving large molecules with all their degrees of vibrational and rotational freedom active it does predict that the a reaction should occur 10^{-9} more slowly than one involving atoms. This calculation is performed just as was shown in Chapter 4 (see Problems 4.6 and 4.7).

We see that both basic theories of chemical reaction may be used in tandem to interpret results. However, more insight can be gained from more detailed use of the modern, much more direct, experimental results, and we move first to considerations of the role of energy in chemical reactions.

5.4   ENERGY CONSIDERATIONS

5.4.1   Activation Energy

Activation energies of chemical reactions are measured experimentally by observing how their rate constants vary with temperature, using the Arrhenius equation (Equation 4.49). A commonly stated reason why a reaction proceeds faster as the temperature is increased is that the fraction of molecules having energy greater than the activation energy increases due to the Maxwell distribution (Figure 1.1) moving to higher average energies and having a broader distribution. This distribution, though, applies to translational motion only. The rationalisation might be correct, and is in some cases, but this conventional explanation does not take into account any possible effect of the internal forms of energy possessed by molecules.

Since values of activation energies are known from normal laboratory experiments we can test the hypothesis that sufficient energy can be obtained simply from the collisions of fast-moving molecules directly in a crossed molecular beam experiment. This is done by varying the velocity of the molecules, using velocity selectors, to study the effects of collisions whose kinetic energy is less than, equal to or greater than E_A. One reaction that has been investigated is that between K atoms and HCl molecules, and it is indeed observed that no reaction occurs until the collision energy exceeds the known activation energy. But it also transpires that the reaction cross section observed is about 100 times lower than that deduced from the normal laboratory experiment – the reaction proceeds, but comparatively inefficiently. This suggests that in this case collisional energy alone is rather ineffective in causing reaction. To see directly whether the internal energy of molecules affects the observations, a laser beam may be used to excite the HCl molecules in one beam from the $v = 0$ ground vibrational state to $v = 1$ – immediately the cross-section climbs and becomes consistent with the laboratory value. This provides direct evidence that vibrational excitation can affect the reaction probability in this reaction.

At first sight this is not surprising. Promoting the molecules to the higher vibrational level implies that the activation energy to reach the transition state is decreased, and the reaction should indeed proceed faster. But with both the activation energy of the reaction and the vibrational energy of the molecule known (the latter from infrared spectroscopy) it is simple to calculate how much faster the reaction should be, and this possible explanation does not account in this reaction for the magnitude of the effect seen. The effect of vibration is more subtle, as will be seen below.

A related experiment pre-dated this one and provided the first evidence that internal energies may have an important role to play in reactions. Studies of the reaction between N and O atoms to form NO molecules exhibited infrared chemiluminescence from the product. This showed that the NO was formed in a highly excited vibrational state, rather than in the $v = 0$ state, and it emitted photons of infrared energy as it dropped down to its lower vibrational and rotational levels. Whereas the K + HCl reaction required vibrational activation of the reactants, this one produced vibrational energy in the products. These effects involving vibrational energy raise a more general energy consideration.

5.4.2 Disposal of Energy and Energy Distribution in Molecules

The driving force for any chemical reaction is that the system lowers its energy from that of two separated reactants by forming new bonds. The potential energy diagram of a pair of H atoms shows a deep minimum when the bond length is reached as the atoms approach. At room temperature the depth of this well is the heat of formation of the H_2 molecule, and is, therefore, large. But there is a problem. It is all very well saying that the energy is lowered, but energy is conserved, and the energy has to be removed or else at the first vibration after the molecule is formed it will dissociate, since it contains precisely the energy needed to do this. The curious fact therefore is that two H atoms encountering in a vacuum cannot react. However, in a gas this changes, since other atoms collide with the molecule as it forms and transport some of the energy away. We conclude that a 'third body' is needed in the collision for reaction between atoms to occur.

Once the molecule is formed it is still subject to collisions from other atoms and molecules in the system. For it to survive these its energy must be dropped by more than kT by the third body. This implies that the total heat of formation need not be removed by the third body for the molecule to be stable, but if only part of it is then the molecule is left in an excited vibrational state. This is what happens in the N + O reaction.

Different 'third body' molecules may be very different in their ability to accept energy and stabilise the product molecule. Instinctively we would expect that a large floppy molecule with many internal degrees of freedom would be more efficient at taking away energy than an atom. The atom can only accept energy as translation, and disappear rapidly into the distance, whereas the molecule might absorb energy into its rotational and vibrational states as well. Study of the energy transfer between molecules is an important research field in its own right, and the condition for efficient transfer turns out to be that the third body should have energy levels that match the energy of the system trying to lose energy. In detail, if the molecule has a high 'density of states' energy transfer is efficient; this is defined as the number of states per cm^{-1} and may reach many millions in large molecules. Obviously, the higher the number the better is the chance of exact energy matching.

Energy transfer is one factor that affects the value of the transmission coefficient in Transition State theory which is, for example, zero for the reaction of two H atoms in a vacuum.

A related, and just as important, problem is that of energy distribution between the different bonds in a molecule. In general a bond dissociates if the energy in it exceeds its dissociation energy. But in our normal concept of activation energy we think of the molecule dissociating when the activation energy is exceeded in the whole molecule. The difference is that a molecule may contain sufficient energy for a specific bond to dissociate but the activation may have occurred in a region of the molecule separate from where the bond breaks. For example, in a photochemical reaction the chromophore that absorbs the energy from a light beam might be well removed in the molecule from the bond that breaks. If this is so then the energy has to be redistributed throughout the molecule until the crucial amount appears in the specific bond. This occurs through the coupled motions of the anharmonic vibrations of the different bonds. The molecule that contains sufficient energy for reaction to occur is said to be activated but it proceeds to reaction only when it becomes possible to traverse the Transition State as the energy distribution (and geometry) becomes appropriate for this.

5.5 POTENTIAL ENERGY SURFACES

We have seen that a wide range of phenomena affect reaction rates and not all are happily encompassed by the simple Transition State and Collision theories that have been described. It is time to take a more fundamental approach.

In describing molecule formation between two atoms we have invoked a two-dimensional potential energy curve that shows how the potential energy varies as one approaches the other, there being just one distance involved. However, even in the simple atom abstraction reaction of a H atom with a H_2 molecule two distances are involved as the reactants approach and then the products separate, r_{AB} and r_{BC}:

$$H_A + H_B - H_C \rightarrow H_A - H_B + H_C$$

where we have labelled the atoms to keep track of them. We need

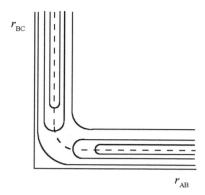

Figure 5.7 *Contour plot of the three-dimensional potential energy surface for the reaction between H atoms and H_2 molecules. As the atom H_A approaches the H_B–H_C molecule (its trajectory is shown by the hatched line) it selects the lowest energy path along the input energy valley. The bond length does not change until it gets very close to the molecule when the reaction proceeds over the local high energy col between the input and exit valleys. The product H_C atom then exits along the bottom of the exit valley. The height of the col above the valley represents the activation energy of the reaction. This reaction is unusual in being symmetric, the reactants and products being identical species.*

therefore to construct a three-dimensional surface to show how the potential energy varies with these two distances. This is difficult to draw and it is conventional to depict such surfaces as contour diagrams, as one would when drawing a geographical map. Figure 5.7 shows that for this reaction.

It is unusual in being completely symmetric about a line at 45° through the origin. It shows two deep valleys joined on this line through a local maximum known as a 'col', maintaining the geographic analogy. As the H_A atom approaches the molecule it travels along the bottom of the valley which then directs it over the col, allowing the reaction to take the lowest energy path. The configuration at the top of the col is the Transition State, and on passing through it reaction has occurred so that atom H_C exits along the other valley. The height of the col above the valleys represents the activation energy. The reaction co-ordinate drawn commonly to depict the course of the reaction is simply the trajectory shown, straightened out and presented in two dimensions.

A more common situation is of an un-symmetric reaction, and an un-symmetric barrier. For the generalised atom/diatomic molecule reaction

$$A + BC \rightarrow AB + C$$

possible potential energy surfaces are shown, again in contour form, in

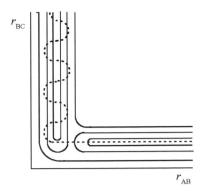

Figure 5.8 *Potential energy surface for a reaction with an early barrier. The reactant approaches from the right along the bottom of the energy valley, goes over the lowest energy point on the surface, the col, and then the trajectory drops into the exit valley. But it enters this valley high up in its walls and, consequently, swings from side to side as it exits, implying vibration produced in the AB bond. That is, the product is formed in an excited vibrational state. An example of this behaviour is the reaction between N and O atoms to produce vibrationally-excited NO.*

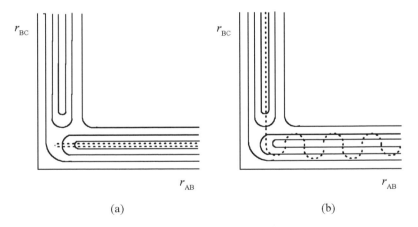

(a)   (b)

Figure 5.9 *Potential energy surfaces with a late barrier to reaction. In* (a) *a reactant enters from the right with just translational energy and runs into a steeply rising energy barrier before it is able to find the lowest energy route, over the col, for reaction. It is simply reflected by this barrier and little or no reaction ensues. However, if, as in* (b)*, the reactant possesses vibrational energy above the ground state it may be able to seek out the lowest energy path and the trajectory traverses the col, as shown in the diagram, to yield products with translational energy only.*

Figures 5.8 and 5.9. They are of two types, one with an early energy col, that is one that occurs before the 45° line, and one with a late col, that occurs after this line. In the 'early barrier' case, one reactant approaches the other as before, the system goes over the col through the transition state and the product exits along the second valley. But because of the unsymmetric position of the col it enters this valley high up on one side of it and the trajectory proceeds to swing from side to side as the product exits. With reference to the axis system, this shows that the bond length of the molecule formed varies regularly with time. That is, the product is formed in a vibrationally excited state. This is the situation in the N + O reaction (which is of course an *atomic* reaction example, but the principle is the same).

The late barrier case is more subtle. Here a reactant travelling along the bottom of the inlet valley with just translational energy essentially runs into a brick wall and is simply reflected out again, with no reaction having occurred. There is a small probability, however, that since the wall is itself asymmetric, sloping slightly towards the col, some reactants are deflected and find the lowest energy path and lead to reaction. However, if the reactant is vibrationally excited the trajectory as it approaches the other reactant has a component in the direction of the col, and this allows it to find the low-energy path more efficiently. This is the behaviour exhibited by the K + HCl reaction and we see very clearly how the reaction efficiency is enhanced due to the existence of just one quantum of vibrational energy, largely independently of any effect on the activation energy which, as mentioned above, is not significantly affected.

The reverse reaction H + KCl would of course provide an example of an early barrier and lead to a vibrationally excited product.

For the Cs + RbCl reaction the potential energy surface is more complicated since a local minimum of energy exists, high above the valleys, corresponding to the tri-atomic collision complex. It is separated from the reactants by one col (transition state) and the products by a second, but the principles remain unaltered.

5.6 SUMMARY

Experiments involving observations of individual encounter processes, and of energy distribution between vibrational modes in products have led to a much more detailed insight into how reactions occur than is possible to obtain using either Collision or Transition State theory as normally applied. The transition state itself has, however, been shown to be a real entity. A complete understanding requires detailed knowledge of the potential energy surfaces, and the rates of simple reactions can now be fully calculated using them. This is beyond the scope of this chapter, whose purpose has been to demonstrate that the modern approach to reaction dynamics is indeed through the properties of primary processes involving atoms and molecules, rather than historical experiments performed on large ensembles of them.

PROBLEMS

Any constants required in the calculations are listed at the end of the book.

5.1   Two different reactions occur with activation energies of (i) 10 and (ii) 100 kJ mol⁻¹. What fraction of molecular collisions occurs in each case with an energy exceeding the activation energy at 300, 600 and 1000 K?

5.2   The collision density, the number of collisions per unit time per unit volume, in a pure gas A is given by

$$Z_{AA} = \sigma_{\text{collision}} \left(\frac{4kT}{\pi m} \right)^{1/2} N_A^2 [A]^2 \; s^{-1} \; m^{-3}$$

where [A] is the molar concentration of A. [This equation differs from the pre-exponential factor in Equation (5.1) that has been converted into molar units. Also a factor of $\frac{1}{2}$ is included so as not to count collisions between *identical* molecules twice and affects the contents of the brackets.]

Evaluate this for molar argon at 298 K given that the argon atom has a collision cross-section of 36×10^{-20} m⁻².

[This yields a very large number. If the calculation is performed for reactive molecules this implies that a significant number of collisions might lead to reaction, even if the fraction with sufficient energy (Problem 5.1) is low.]

5.3. For the reaction

$$NO + O_3 \rightarrow NO_2 + O_2$$

the rate constant is 6.3×10^7 dm³ mol⁻¹s⁻¹ at 500 K and 16.4×10^7 dm³ mol⁻¹ s⁻¹ at 800 K. What is the pre-exponential factor A and the activation energy for this reaction? (Assume A to be independent of temperature.)

Using the collision diameters of the two reactants, A is calculated to be 3.98×10^8 dm³ mol⁻¹ s⁻¹. What is the value of the steric factor P?

5.4   The rate constant for the reaction

$$NOCl + NOCl \rightarrow 2NO + Cl$$

is 4.418×10^4 dm³ mol⁻¹ s⁻¹ at 1000 K and the activation energy is 102.0 kJ mol⁻¹. What is the reaction cross-section?

(Take the relative atomic masses of N,O and Cl to be 14,16 and 35.5 respectively.)

5.5 In the reaction of two methyl radicals to form ethane the pre-exponential factor $A = 2.4 \times 10^{10}$ dm^3 mol^{-1} s^{-1}. Calculate the reaction cross-section.

The methyl radical, $CH_3^{\bullet}$, is planar but sweeps out a cylinder as it travels through space as a result of rotation. Methane, CH_4, has a collision cross-section of 46×10^{-20} m^2. Assume that this is the collision cross-section of the radical, too, and work out an approximate value for P.

Answers to problems

Chapter 1

1.1 (i) $6.5RT$, $6.5R$; (ii) $6RT$, $6R$.

1.2 (i) $n_1 = 3.010 \times 10^{23}$, $n_0 = 3.012 \times 10^{23}$; (ii) 6.832×10^{-5} J;
(iii) 4.88×10^{-6} J T^{-1}
(iv) 515 MHz.

1.4 (i) $0.067\,N_A$, 1.62×10^{25} cm^{-1} mol^{-1} $\equiv 322.1$ J mol^{-1}
(ii) $0.195\,N_A$, 1.261×10^{26} cm^{-1} mol^{-1} $\equiv 2504.3$ J mol^{-1}
(iii) $0.243\,N_A$, 2.553×10^{26} cm^{-1} mol^{-1} $\equiv 5070.3$ J mol^{-1}.

1.5 (i) 0.014, 3.694×10^{24} cm^{-1} mol^{-1} $\equiv 7.3$ J mol^{-1}
(ii) 1.889×10^{-8}, 4.995×10^{18} cm^{-1} mol^{-1} $\equiv 9.92 \times 10^{-3}$ J mol^{-1}.

1.6 degeneracies of 5,3,1; $^2P_1$: $0.208\,N_A$, $^2P_0$: $0.150\,N_A$, $^2P_2$ $0.742\,N_A$;
2.667×10^{25} cm^{-1} mol^{-1} $\equiv 529.67$ J mol^{-1}.

1.7 $0.648\,N_A$, $0.602\,N_A$, $0.241\,N_A$, $0.065\,N_A$; 2132 cm^{-1}.

Chapter 2

2.1 10–11, 19.90 (calculated to 14 terms). No electronic contribution.

2.2 19.6. Worse for lighter molecule.

2.3 $B = 0.2457$ cm^{-1}, $q_{rot} = 421.7$.

2.4 19.93, in contrast to 19.87. Difference largely negligible in calculating thermodynamic properties.

2.5 $\theta_{rot} = 15.25$, 0.75, 0.16 K. Rotational energy becomes significant at low temperatures, and at particularly low ones for the heavier molecules.
$\theta_{vib} = 4156$ K for H^{35}Cl, 1128 K for F^{35}Cl. Again the temperatures are lower for heavier molecules.

2.6 At 298 K 2.09×10^{28}, at 0.01 K 4.16×10^{21}.

2.7 4.09×10^{29}.

Chapter 3

3.1 Answer in question.

3.2 $\Delta S = A[\ln T]_{T_1}^{T_2} + B[T]_{T_1}^{T_2} + \frac{1}{2}C[T^2]_{T_1}^{T_2}$; 1326.55 J K^{-1} mol^{-1}.

3.3 N_2: 11.37, 72.1 J K^{-1} mol^{-1}; NH_3: 28.92, 97.0 J K^{-1} mol^{-1}; H_2O: 22.0, 109.0 J K^{-1} mol^{-1}

3.4 191.2 J K^{-1} mol^{-1}.

3.5 Entropy of monoclinic form at 369 K = entropy of rhombic form at 369 K + the entropy of the phase change if the Third Law is obeyed. It is.

3.6 577.9 JK^{-1}mol^{-1}. For ^{16}O–^{16}O, $S(0) = 0$; for ^{18}O–^{16}O $S(0) \sim R\ln 2$.

Chapter 4

4.1 Sackur–Tetrode values: 126, 146.4, 154.8, 164.0 and 169.7 J K^{-1} mol^{-1}. Close agreement with experiment.

4.2 At 298 K we expect a further contribution from rotation for HF, but not vibration.

$q_{rot} = 9.63$. $S_{rot} = R$ (from Equipartition) + $R\ln 9.63 = 27.14$. This is the difference observed.

For H_2O, which is non-linear, we expect an additional rotational contribution, and a possible small contribution from lower-lying vibrational levels.

4.3 0.0063 J mol^{-1} T^{-2}; 0.63 J mol^{-1}; 627 T (presently unattainable).

4.4 *Ortho* and *para* molecules, according to the Pauli principle.

4.5 $K_P = 3.14 \times 10^{-11}$.

4.6 (i) $k = \kappa \dfrac{kT}{h} \dfrac{RT}{P^{\circ}} q_{TS,rot}^2 \dfrac{\left[\dfrac{2\pi(m+n)kT}{h^2}\right]^{3/2}}{\left(\dfrac{2\pi mkT}{h^2}\right)^{3/2}\left(\dfrac{2\pi nkT}{h^2}\right)^{3/2}} e^{-E_A/RT}$

(ii) $k \approx \kappa \dfrac{kT}{h} \dfrac{RT}{P^{\circ}} q_{TS,rot}^2 \dfrac{1}{\left(\dfrac{2\pi mkT}{h^2}\right)^{3/2}} e^{-E_A/RT}$

4.7 For reaction between complex molecules,

$k \approx \kappa \dfrac{kT}{h} \dfrac{RT}{P^{\circ}} \dfrac{q_{vib}^5}{q_{rot}^3} \dfrac{1}{\left(\dfrac{2\pi mkT}{h^2}\right)^{3/2}} e^{-E_A/RT}$

Assuming the same values for κ and for E_A then the ratio of this to the answer to Problem 6(ii) is q_{vib}^5/q_{rot}^5, about 3×10^{-9}.

Chapter 5

5.1   (i) 0.018, 0.135, 0.300; (ii) 3.8×10^{-18}, 2.0×10^{-9}, 6.0×10^{-6}.

5.2 1.16×10^{30} $s^{-1}m^{-3}$.

5.3 $A = 7.94 \times 10^8$ dm^3 $mol^{-1}s^{-1}$, 10.5 kJ mol^{-1}, $P = 2$.

5.4 0.36×10^{-20} m^2.

5.5 13.7×10^{-20} m^2; 0.3.

Some Useful Constants and Relations

Boltzmann constant $\qquad\qquad k = 1.381 \times 10^{-23}$ J K^{-1}.

Avogadro number $\qquad\qquad N_A = 6.022 \times 10^{23}$ mol^{-1}.

Gas constant $\qquad\qquad R = kN_A = 8.314$ J K^{-1} mol^{-1}.

Planck's constant $\qquad\qquad h = 6.63 \times 10^{-34}$ J s.

Thermal energy $\qquad\qquad kT = 4.116 \times 10^{-21}$ J at 298 K.
$\qquad\qquad\qquad\qquad\qquad kT/hc = 207.2$ cm^{-1} at 298 K.

$hc/k = 1.44$ cmK.
1 cm$^{-1} \equiv 1.986 \times 10^{-23}$ J.

Bohr magneton $\qquad\qquad \mu_B = 9.27 \times 10^{-24}$ J T^{-1}.

g factor for the free electron $\qquad g_e = 2.0023$.

Velocity of light $\qquad\qquad c = 2.998 \times 10^8$ m s^{-1}.

Standard molar volume $\qquad V_m^{\circ} = RT/P^{\circ} = 2.479 \times 10^{-2}$ m^3 mol^{-1}, where $T = 298$ K.

Standard pressure $\qquad\qquad P^{\circ} = 100$ kPa (Pascal) $= 10^5$ N m^{-2}.

Atmospheric pressure $\qquad 1$ atm $= 101.35$ kPa (by definition).

Further Reading

Some diverse general texts that cover the contents in this book are:

P.W. Atkins and J. de Paula, *Physical Chemistry*, Oxford University Press, 2002

S.S. Berry, S.A. Rice and J. Ross, *Physical Chemistry*, Oxford University Press, 2000

D.A. McQuarrie and J.D. Simon, *Physical Chemistry: A Molecular Approach*, University Science Books, 1997.

On more specialist topics the reader will find the following useful:

M.J. Pilling and P.W. Seakins, *Reaction Kinetics*, Oxford University Press, 1995

S.R. Logan, *Chemical Kinetics*, Mangum, Harlow, 1996

J.I. Steinfeld, J.S. Francisco and W.L. Hase, *Chemical Kinetics and Dynamics*, *Prentice Hall, Englewood Cliffs*, 1998.

C.N. Banwell and E.M. McCash, *Fundamentals of Molecular Spectroscopy*, McGraw-Hill, 1994.

Subject Index